DESIGN OF AMPLIFIERS
AND OSCILLATORS
BY THE S-PARAMETER
METHOD

DESIGN OF AMPLIFIERS
AND OSCILLATORS
BY THE S-PARAMETER
METHOD

GEORGE D. VENDELIN
Technical Director
Electronic Instrumentation Division
Eaton Corporation
Sunnyvale, California

A WILEY-INTERSCIENCE PUBLICATION

JOHN WILEY & SONS

New York   Chichester   Brisbane   Toronto   Singapore

Library of Congress Cataloging in Publication Data:

Vendelin, George D. (George David), 1938-
    Design of amplifiers and oscillators by the S-parameter method.

    "A Wiley-Interscience publication."
    Includes bibliographies and index.
    1. Microwave integrated circuits. 2. Microwave
amplifiers 3. Oscillators, Microwave.
4. Electronic circuit design. I. Title.

TK7876.V46.   621.381'325.   81-13005
ISBN 0-471-09226-6   AACR2

Printed in the United States of America

10 9 8 7 6 5 4 3

To the memory of my brother Tommy,
who gave his life for his country,
my wife Barbara,
and my parents, Grace and George

FOREWORD

During the past 15 years, a great number of technical papers have been written and presented on the fundamentals and applications of scattering-parameter design techniques. Most of the early articles focused on the idealized design case, and although some of the later ones covered practical design cases, they were still dedicated to specific applications. Until now nothing has been available in the form of an up-to-date, continuous, and coherent textbook to tie together the various concepts involved. George Vendelin's book approaches the topic from a practical viewpoint and also presents the mathematical derivations and definitions of the key circuit parameters involved. The book also includes a collection of practical design techniques that should be of value in understanding and developing various active and passive microwave circuits. In my opinion, it should provide great help both to students and to practicing microwave engineers in mastering this essential subject matter.

LES BESSER

President
Compact Engineering
Palo Alto, California
August 1981

PREFACE

This book has evolved from my experiences in teaching a graduate-level course on this subject at several universities, including the University of California Extension, UCLA, and the University of Santa Clara. Since an adequate text could not be found, this book has been an ongoing project for several years.

The intention of this book is to prepare the student for effectively using the computer for microwave integrated circuit (MIC) design. While most practicing engineers will quickly find the value of COMPACT or SUPER-COMPACT, there is a need for an introductory textbook to prepare the engineer for the best use of computer-aided design (CAD). This book will not only serve as a useful reference but also should stimulate new ideas and concepts for the creative designer.

I have divided the book into four chapters. The first two chapters cover the basic tools for amplifier and oscillator design. The first chapter includes S-parameter two-port network theory and the numerous definitions of power gain. The second chapter describes microwave transistors, microstriplines, and impedance matching techniques.

The third chapter is devoted to amplifier designs, including the special considerations of balanced amplifier design, high-power design, broadband amplifier design, feedback design, and two-stage amplifier design. The basic limitations in device and circuit performance are discussed for each of these topics.

The fourth chapter deals with oscillator design. The concept of the compressed Smith Chart and the application of S-parameters to oscillator design are covered in this chapter. In addition, the special considerations for low-noise design, high-power design, broadband design, and buffered oscillator design are included in this chapter.

Several colleagues have contributed to the information in this book, but I must especially thank the following: Dr. Behruz Rezvani, who reviewed Chapters 1 and 4; Bill Chan, who rewrote the section on low-noise oscillator design; Dr. Al Sweet, who edited Chapter 4; and Dr. Steve Ludvik, who reviewed the entire book and made many helpful suggestions. Some of the essential principles in this book were explained to me by Dr. Yozo Satoda of Dexcel, Inc., and Les Besser of Compact Engineering. I also wish to thank Dr. Gunther Sorger for providing the proper working conditions for the completion of this project and Mrs. Patty Brown for her careful typing of the manuscript. Finally, I thank my wife Barbara for giving me the time.

GEORGE D. VENDELIN

Sunnyvale, California
November 1981

CONTENTS

DESIGN OF AMPLIFIERS
AND OSCILLATORS
BY THE S-PARAMETER
METHOD

CHAPTER ONE

S-PARAMETERS

1.0   Introduction

The necessary tools for the design of microwave amplifiers and oscillators are found from an understanding of transmission lines, two-port networks, and impedance matching techniques. With a knowledge of the transistor S-parameters, the design of amplifier and oscillator circuits is essentially an impedance matching problem.

The amplifier and oscillator design techniques are shown schematically in Fig. 1.1. For the amplifier, the power available from the source should be delivered to the input of the transistor; this is the input matching problem M_1, which is an idealized lossless matching circuit. The output lossless matching structure M_2 (also an idealized lossless matching circuit) should be designed to deliver the maximum power to the load. This type of amplifier, which is simultaneously conjugately matched at input and output ports, is only possible when the stability factor is sufficiently large ($k > 1$). The power gain is the maximum available gain (G_{ma}).

The oscillator is a similar design problem, where now the stability factor must be less than one ($k < 1$). Notice that the load is receiving power for either case, that is, the load would not know whether the transistor is an amplifier or an oscillator. Usually the same transistors are used for both applications. Some type of feedback may be required to bring $k < 1$ at the frequency of interest. The lossless input structure M_3 resonates the input port, whereas the output matching structure M_4 should be designed to deliver the maximum power to the load.

1.1   Transmission Lines

Electrical transmission lines are used for propagation of electromagnetic waves. Usually one conductor is the signal side and other is ground, that is, the transmission line is unbalanced. Examples of common TEM (transverse-electric-magnetic fields) transmission lines are shown in Fig. 1.2, where the signal side is the inner conductor of the coaxial cable, the inner conductor of the stripline, and the top conductor of the microstripline.

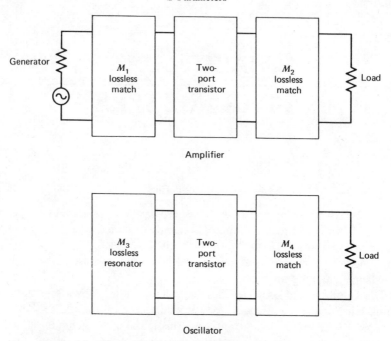

Figure 1.1 Block diagram for amplifier and oscillator design.

The microstrip case is popular for microwave integrated circuits, since the signal line (or center conductor) is readily accessible for interconnection of active components. This transmission line is described in detail in Section 2.4. A waveguide transmission line is not compatible with active component design, since there is no center conductor for signal connection. Moreover, this transmission line propagates a TE_{10} mode, which unnecessarily restricts the bandwidth of the circuit because of dispersion of the signal.

Transmission lines are characterized by three parameters: the characteristic impedance Z_0, the phase velocity v_{ph}, and the attenuation constant α. The electrical behavior of transmission lines (see Fig. 1.3) is described by a set of four traveling waves (where only two waves are independent) with rms values given by:

$$V_{inc} \qquad \text{Incident voltage wave}$$

$$I_{inc} = \frac{V_{inc}}{Z_0} \qquad \text{Incident current wave}$$

$$V_{ref} \qquad \text{Reflected voltage wave}$$

$$I_{ref} = \frac{V_{ref}}{Z_0} \qquad \text{Reflected current wave}$$

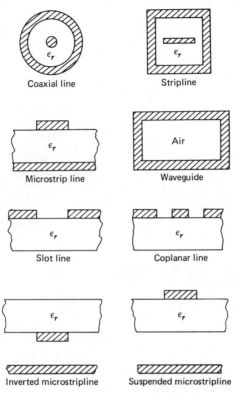

Figure 1.2 Common transmission lines.

Each wave is a sinusoidal voltage with time and distance variation given by

$$\sqrt{2}\, V_{\text{inc}} e^{j\omega t} e^{-\gamma x} \qquad \text{A voltage wave traveling in the } +x \text{ direction}$$

$$\sqrt{2}\, I_{\text{inc}} e^{j\omega t} e^{-\gamma x} \qquad \text{A current wave traveling in the } +x \text{ direction}$$

$$\sqrt{2}\, V_{\text{ref}} e^{j\omega t} e^{\gamma x} \qquad \text{A voltage wave traveling in the } -x \text{ direction}$$

$$\sqrt{2}\, I_{\text{ref}} e^{j\omega t} e^{\gamma x} \qquad \text{A current wave traveling in the } -x \text{ direction}$$

Here γ is the propagation constant defined by

$$\gamma = \alpha + j\beta \tag{1.1}$$

where α is the attenuation factor in nepers per meter and β is the phase constant in radians per meter. Normally the time dependence is dropped for simplicity. Thus the total voltage and current at any point on the line is

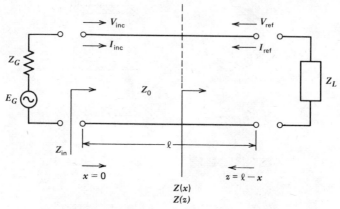

Figure 1.3   Traveling waves on a transmission line.

given by

$$V(x) = \sqrt{2}\, V_{\text{inc}} e^{-\gamma x} + \sqrt{2}\, V_{\text{ref}} e^{\gamma x} \qquad (1.2)$$

$$I(x) = \sqrt{2}\, I_{\text{inc}} e^{-\gamma x} - \sqrt{2}\, I_{\text{ref}} e^{\gamma x} \qquad (1.3)$$

The boundary conditions for the transmission line are required to complete the analysis. At the generator a convenient (and practical) boundary condition is $Z_G = Z_0$. At the load a reflection coefficient is defined by

$$\Gamma_L = \frac{V_{\text{ref}} e^{\gamma \ell}}{V_{\text{inc}} e^{-\gamma \ell}} = \frac{V_{\text{ref}}}{V_{\text{inc}}} e^{2\gamma \ell} \qquad (1.4)$$

If $Z_L = Z_0$, $\Gamma_L = V_{\text{ref}} = 0$. Since

$$Z_L = \frac{V(\ell)}{I(\ell)} = \frac{V_{\text{inc}} e^{-\gamma \ell} + V_{\text{ref}} e^{\gamma \ell}}{I_{\text{inc}} e^{-\gamma \ell} - I_{\text{ref}} e^{\gamma \ell}} \qquad (1.5)$$

$$Z_L = Z_0 \frac{V_{\text{inc}} e^{-\gamma \ell} + V_{\text{ref}} e^{\gamma \ell}}{V_{\text{inc}} e^{-\gamma \ell} - V_{\text{ref}} e^{\gamma \ell}} = Z_0 \frac{1 + \Gamma_L}{1 - \Gamma_L} \qquad (1.6)$$

$$\Gamma_L = \frac{Z_L - Z_0}{Z_L + Z_0} \qquad (1.7)$$

The input impedance at any point on the line is given by

$$Z(x) = \frac{V(x)}{I(x)} = Z_0 \frac{V_{\text{inc}} e^{-\gamma x} + V_{\text{ref}} e^{\gamma x}}{V_{\text{inc}} e^{-\gamma x} - V_{\text{ref}} e^{\gamma x}} \qquad (1.8)$$

Let $z = \ell - x$.

$$Z(z) = Z_0 \frac{1 + \Gamma_L e^{-2\gamma z}}{1 - \Gamma_L e^{-2\gamma z}} \tag{1.9}$$

Comparison of (1.9) with (1.6) shows that the input impedance is a load impedance with a new reflection coefficient of $\Gamma_L e^{-2\gamma z}$, that is, with a phase rotation of $e^{-2\gamma z}$. At $x = 0$,

$$Z_{in} = Z_0 \frac{1 + \Gamma_L e^{-2\gamma \ell}}{1 - \Gamma_L e^{-2\gamma \ell}} \tag{1.10}$$

Since

$$\Gamma_{in} = \Gamma_L e^{-2\gamma \ell} \tag{1.11}$$

$$Z_{in} = Z_0 \frac{1 + \Gamma_{in}}{1 - \Gamma_{in}} \tag{1.12}$$

Therefore

$$\Gamma_{in} = \frac{Z_{in} - Z_0}{Z_{in} + Z_0} \tag{1.13}$$

The attenuation constant α decreases the voltage and current as the wave travels along the line. Usually this effect can be ignored in well-designed microwave integrated circuits. The phase velocity is related to β by

$$\beta = \frac{\omega}{v_{ph}} = \frac{2\pi f}{v_{ph}} = \frac{2\pi}{\lambda_g} \tag{1.14}$$

Therefore for movement of $j\beta l = 2\pi$, the wave has traveled one wavelength or $x = \lambda_g$. For a line with an effective dielectric constant of ε_{eff}, phase ? velocity is decreased or the wave is slowed to

$$v_{ph} = \frac{c}{\sqrt{\varepsilon_{eff}}} \tag{1.15}$$

where c is the velocity of light. Thus the wavelength (λ_g) is decreased by this type of line, which is a technique for reducing the size of a microwave integrated circuit.

In summary, the voltage and current on a transmission line are functions of time and distance; boundary conditions Z_L and Z_G; and line constants Z_0, v_{ph}, and α. Four traveling waves with the boundary conditions given by the generator and load determine the voltage and current solutions. The load impedance Z_L can be described by a reflection coefficient Γ_L, and the phase of the reflection coefficient changes by $e^{-2\gamma z}$ as one moves from the load to the generator.

1.2 Two-Port Networks

Electrical networks can be described by the number of external terminals that are available for analysis. A single terminal pair is a one-port. A two-port may be formed by either two terminal pairs or two terminals above ground (Fig. 1.4). A three-port may be formed by three terminal pairs or three terminals above ground (e.g., the transistor three-port in Fig. 1.4).

The external behavior of a one-port is described by the voltage and current simply by Ohm's law.

$$Z_1 = \frac{V_1}{I_1} \tag{1.16}$$

If a transmission line Z_0 is connected to the one-port (see Fig. 1.5), the

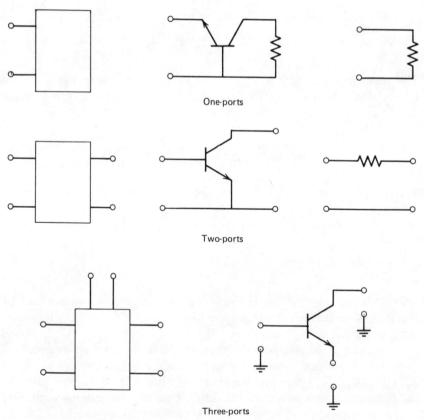

One-ports

Two-ports

Three-ports

Figure 1.4   One-ports, two-ports, and three-ports.

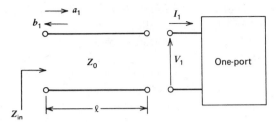

Figure 1.5 Description of a one-port.

previous section has shown that

$$Z_{in} = Z_0 \frac{1 + \Gamma_{in}}{1 - \Gamma_{in}} \tag{1.17}$$

where

$$\Gamma_{in} = \Gamma_L e^{-2\gamma\ell} \tag{1.18}$$

and

$$\Gamma_L = \frac{Z_1 - Z_0}{Z_1 + Z_0} \tag{1.19}$$

The one-port network has a reflection coefficient associated with the input of Γ_{in}. Defining the incident and reflected voltage waves as follows will lead to the definition of S-parameters or scattering parameters:

$$a_1 = \frac{V_{inc}}{\sqrt{Z_0}} \tag{1.20}$$

$$b_1 = \frac{V_{ref}}{\sqrt{Z_0}} \tag{1.21}$$

These new waves are considered power waves, since the dimensions are (watts)$^{1/2}$:

$$\text{incident power} = |a_1|^2 = \frac{|V_{inc}|^2}{Z_0} \tag{1.22}$$

$$\text{reflected power} = |b_1|^2 = \frac{|V_{ref}|^2}{Z_0} \tag{1.23}$$

In terms of the voltage reflection coefficient,

$$\Gamma_{in} = \frac{V_{ref}}{V_{inc}} = \frac{b_1}{a_1} = S \tag{1.24}$$

where S is the scattering parameter of the one-port. This is a complex number that describes the voltage reflection coefficient of the one-port.

In a similar manner the two-port can be described by the external voltages and currents at each port or by the incident and reflected voltages at each port. For the two-port network (Fig. 1.6) we have an input and output reflection coefficient for the total two-port given by:

$$\Gamma_{in} = \Gamma_1 e^{-2\gamma \ell_1} = \frac{b_1}{a_1} \qquad (1.25)$$

$$\Gamma_1 = \frac{Z_1 - Z_{01}}{Z_1 + Z_{01}} \qquad (1.26)$$

$$\Gamma_{out} = \Gamma_2 e^{-2\gamma \ell_2} = \frac{b_2}{a_2} \qquad (1.27)$$

$$\Gamma_2 = \frac{Z_2 - Z_{02}}{Z_2 + Z_{02}} \qquad (1.28)$$

The a and b waves are again the incident and reflected power waves (when squared) and are given by

$$a_1 = \frac{V_{inc1}}{\sqrt{Z_{01}}} \qquad (1.29)$$

$$b_1 = \frac{V_{ref1}}{\sqrt{Z_{01}}} \qquad (1.30)$$

$$a_2 = \frac{V_{inc2}}{\sqrt{Z_{02}}} \qquad (1.31)$$

$$b_2 = \frac{V_{ref2}}{\sqrt{Z_{02}}} \qquad (1.32)$$

The scattering parameters of the two-port are defined by

$$b_1 = S_{11} a_1 + S_{12} a_2 \qquad (1.33)$$

$$b_2 = S_{21} a_1 + S_{22} a_2 \qquad (1.34)$$

It is convenient to consider the a's independent variables and the b's dependent variables. The waves arrive at the two-port in alphabetical order: "a" is incident, "b" is reflected. The reflection coefficients of the two-port

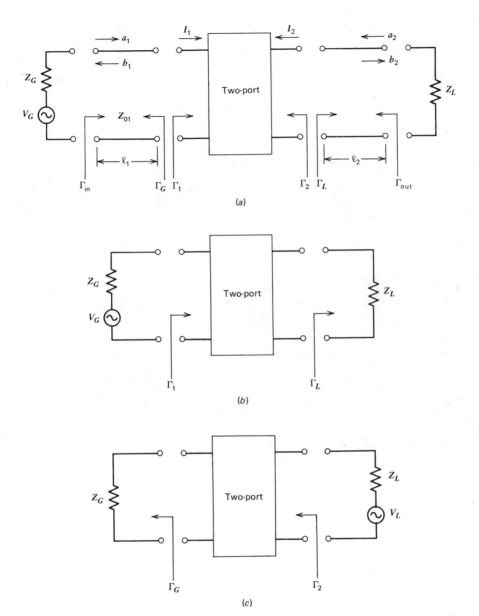

Figure 1.6 Description of a two-port. (*a*) Two-port with generator at port 1 and load at port 2; (*b*) measurement of Γ_1; (*c*) measurement of Γ_2.

9

in Fig. 1.6 can be found by setting $Z_{01} = Z_{02} = Z_0$ and $\ell_1 = \ell_2 = 0$ for convenience. There is only a phase shift associated with the length of the input and output lossless transmission lines, and this can be temporarily dropped. Solving for (1.25) with the help of (1.33), (1.34), and

$$\Gamma_L = \frac{a_2}{b_2} \qquad (1.35)$$

gives

$$\Gamma_1 = \frac{b_1}{a_1} = S_{11} + \frac{S_{12}S_{21}\Gamma_L}{1 - S_{22}\Gamma_L} = S_{11}' \qquad (1.36)$$

Similarly, solving for (1.27) with the help of (1.33), (1.34), and

$$\Gamma_G = \frac{a_1}{b_1} \qquad (1.37)$$

gives

$$\Gamma_2 = \frac{b_2}{a_2} = S_{22} + \frac{S_{12}S_{21}\Gamma_G}{1 - S_{11}\Gamma_G} = S_{22}' \qquad (1.38)$$

When $\Gamma_L = \Gamma_G = 0$, these reduce to

$$\Gamma_1 = S_{11} \qquad (1.39)$$

$$\Gamma_2 = S_{22} \qquad (1.40)$$

The input and output reflection coefficients are functions of all four S-parameters and the terminations Γ_L and Γ_G.

The incident and reflected rms voltage waves are directly related to the voltages and currents at each port in a simple way. At the input terminals of the two-port ($\ell_1 = \ell_2 = 0$),

$$V_1 = V_{inc} + V_{ref} \qquad (1.41)$$

$$Z_0 I_1 = V_{inc} - V_{ref} \qquad (1.42)$$

Adding and subtracting these equations gives

$$V_{inc} = \frac{V_1 + Z_0 I_1}{2} \qquad (1.43)$$

$$V_{ref} = \frac{V_1 - Z_0 I_1}{2} \qquad (1.44)$$

Therefore

$$a_1 = \frac{V_1 + Z_0 I_1}{2\sqrt{Z_0}} \tag{1.45}$$

$$b_1 = \frac{V_1 - Z_0 I_1}{2\sqrt{Z_0}} \tag{1.46}$$

$$a_2 = \frac{V_2 + Z_0 I_2}{2\sqrt{Z_0}} \tag{1.47}$$

$$b_2 = \frac{V_2 - Z_0 I_2}{2\sqrt{Z_0}} \tag{1.48}$$

Since the two-port is often described by other sets of parameters, an equivalence between parameter sets is useful. If any one set of two-port parameters is measured, other sets can be calculated using Table 1.1. The y parameter, z-parameter, $ABCD$, and h-parameter sets are given below.

$$I_1 = y_{11}V_1 + y_{12}V_2 \tag{1.49}$$

$$I_2 = y_{21}V_1 + y_{22}V_2 \tag{1.50}$$

$$V_1 = z_{11}I_1 + z_{12}I_2 \tag{1.51}$$

$$V_2 = z_{21}I_1 + z_{22}I_2 \tag{1.52}$$

$$V_1 = AV_2 - BI_2 \tag{1.53}$$

$$I_1 = CV_2 - DI_2 \tag{1.54}$$

$$V_1 = h_{11}I_1 + h_{12}V_2 \tag{1.55}$$

$$I_2 = h_{21}I_1 + h_{22}V_2 \tag{1.56}$$

The S-parameters have become the most popular for microwave transistors for two reasons: measurement accuracy and transistor stability. Often a transistor may oscillate with a short circuit or open circuit at one port, thus invalidating the measurement of y_{11} or z_{11}.

The extension of this analysis to three-ports or four-ports is straightforward, but the number of parameters becomes very large. However, a

Table 1.1 Conversions Between Two-Port Parameters Normalized to $Z_0 = 1$ with $\Delta^K = K_{11}K_{22} - K_{12}K_{21}$

	S	Z	Y	H	A
S	$\begin{bmatrix} b_1 \\ b_2 \end{bmatrix} = \begin{bmatrix} S_{11} & S_{12} \\ S_{21} & S_{22} \end{bmatrix} \begin{bmatrix} a_1 \\ a_2 \end{bmatrix}$	$S_{11} = \dfrac{(Z_{11}-1)(Z_{22}+1) - Z_{12}Z_{21}}{(Z_{11}+1)(Z_{22}+1) - Z_{12}Z_{21}}$ $S_{12} = \dfrac{2Z_{12}}{(Z_{11}+1)(Z_{22}+1) - Z_{12}Z_{21}}$ $S_{21} = \dfrac{2Z_{21}}{(Z_{11}+1)(Z_{22}+1) - Z_{12}Z_{21}}$ $S_{22} = \dfrac{(Z_{11}+1)(Z_{22}-1) - Z_{12}Z_{21}}{(Z_{11}+1)(Z_{22}+1) - Z_{12}Z_{21}}$	$S_{11} = \dfrac{(1-Y_{11})(1+Y_{22}) + Y_{12}Y_{21}}{(1+Y_{11})(1+Y_{22}) + Y_{12}Y_{21}}$ $S_{12} = \dfrac{-2Y_{12}}{(1+Y_{11})(1+Y_{22}) - Y_{12}Y_{21}}$ $S_{21} = \dfrac{-2Y_{21}}{(1+Y_{11})(1+Y_{22}) - Y_{12}Y_{21}}$ $S_{22} = \dfrac{(1+Y_{11})(1-Y_{22}) + Y_{12}Y_{21}}{(1+Y_{11})(1+Y_{22}) - Y_{12}Y_{21}}$	$S_{11} = \dfrac{(h_{11}-1)(h_{22}+1) - h_{12}h_{21}}{(h_{11}+1)(h_{22}+1) - h_{12}h_{21}}$ $S_{12} = \dfrac{2h_{12}}{(h_{11}+1)(h_{22}+1) - h_{12}h_{21}}$ $S_{21} = \dfrac{-2h_{21}}{(h_{11}+1)(h_{22}+1) - h_{12}h_{21}}$ $S_{22} = \dfrac{(1+h_{11})(1-h_{22}) + h_{12}h_{21}}{(h_{11}+1)(h_{22}+1) - h_{12}h_{21}}$	$\dfrac{A+B-C-D}{A+B+C+D} \quad \dfrac{2(AD-BC)}{A+B+C+D}$ $\dfrac{2}{A+B+C+D} \quad \dfrac{-A+B-C+D}{A+B+C+D}$
Z	$Z_{11} = \dfrac{(1+S_{11})(1-S_{22}) + S_{12}S_{21}}{(1-S_{11})(1-S_{22}) - S_{12}S_{21}}$ $Z_{12} = \dfrac{2S_{12}}{(1-S_{11})(1-S_{22}) - S_{12}S_{21}}$ $Z_{21} = \dfrac{2S_{21}}{(1-S_{11})(1-S_{22}) - S_{12}S_{21}}$ $Z_{22} = \dfrac{(1-S_{11})(1+S_{22}) + S_{12}S_{21}}{(1-S_{11})(1-S_{22}) - S_{12}S_{21}}$	$\begin{bmatrix} V_1 \\ V_2 \end{bmatrix} = \begin{bmatrix} Z_{11} & Z_{12} \\ Z_{21} & Z_{22} \end{bmatrix} \begin{bmatrix} I_1 \\ I_2 \end{bmatrix}$	$\dfrac{Y_{22}}{\Delta^Y} \quad \dfrac{-Y_{12}}{\Delta^Y}$ $\dfrac{-Y_{21}}{\Delta^Y} \quad \dfrac{Y_{11}}{\Delta^Y}$	$\dfrac{\Delta^h}{h_{22}} \quad \dfrac{h_{12}}{h_{22}}$ $\dfrac{-h_{21}}{h_{22}} \quad \dfrac{1}{h_{22}}$	$\dfrac{A}{C} \quad \dfrac{\Delta^A}{C}$ $\dfrac{1}{C} \quad \dfrac{D}{C}$
Y	$Y_{11} = \dfrac{(1-S_{11})(1+S_{22}) + S_{12}S_{21}}{(1+S_{11})(1+S_{22}) - S_{12}S_{21}}$ $Y_{12} = \dfrac{-2S_{12}}{(1+S_{11})(1+S_{22}) - S_{12}S_{21}}$ $Y_{21} = \dfrac{-2S_{21}}{(1+S_{11})(1+S_{22}) - S_{12}S_{21}}$ $Y_{22} = \dfrac{(1+S_{11})(1-S_{22}) + S_{12}S_{21}}{(1+S_{11})(1+S_{22}) - S_{12}S_{21}}$	$\dfrac{Z_{22}}{\Delta^Z} \quad \dfrac{-Z_{12}}{\Delta^Z}$ $\dfrac{-Z_{21}}{\Delta^Z} \quad \dfrac{Z_{11}}{\Delta^Z}$	$\begin{bmatrix} I_1 \\ I_2 \end{bmatrix} = \begin{bmatrix} Y_{11} & Y_{12} \\ Y_{21} & Y_{22} \end{bmatrix} \begin{bmatrix} V_1 \\ V_2 \end{bmatrix}$	$\dfrac{1}{h_{11}} \quad \dfrac{-h_{12}}{h_{11}}$ $\dfrac{h_{21}}{h_{11}} \quad \dfrac{\Delta^h}{h_{11}}$	$\dfrac{D}{B} \quad \dfrac{-\Delta^A}{B}$ $\dfrac{-1}{B} \quad \dfrac{A}{B}$

	S-parameters	Z-parameters	Y-parameters	h-parameters	A-parameters
H	$h_{11} = \dfrac{(1+S_{11})(1+S_{22})-S_{12}S_{21}}{(1-S_{11})(1+S_{22})+S_{12}S_{21}}$ $h_{12} = \dfrac{2S_{12}}{(1-S_{11})(1+S_{22})+S_{12}S_{21}}$ $h_{21} = \dfrac{-2S_{21}}{(1-S_{11})(1+S_{22})+S_{12}S_{21}}$ $h_{22} = \dfrac{(1-S_{22})(1-S_{11})-S_{12}S_{21}}{(1-S_{11})(1+S_{22})+S_{12}S_{21}}$	$\begin{bmatrix} \dfrac{\Delta^Z}{Z_{22}} & \dfrac{Z_{12}}{Z_{22}} \\[2mm] \dfrac{-Z_{21}}{Z_{22}} & \dfrac{1}{Z_{22}} \end{bmatrix}$	$\begin{bmatrix} \dfrac{1}{Y_{11}} & \dfrac{-Y_{12}}{Y_{11}} \\[2mm] \dfrac{Y_{21}}{Y_{11}} & \dfrac{\Delta^Y}{Y_{11}} \end{bmatrix}$	$\begin{bmatrix} V_1 \\ I_2 \end{bmatrix} = \begin{bmatrix} h_{11} & h_{12} \\ h_{21} & h_{22} \end{bmatrix}\begin{bmatrix} I_1 \\ V_2 \end{bmatrix}$	$\begin{bmatrix} \dfrac{\Delta^A}{D} & \dfrac{B}{D} \\[2mm] \dfrac{C}{D} & \dfrac{-1}{D} \end{bmatrix}$
A	$A = \dfrac{(1+S_{11})(1-S_{22})+S_{12}S_{21}}{2S_{21}}$ $B = \dfrac{(1+S_{11})(1+S_{22})-S_{12}S_{21}}{2S_{21}}$ $C = \dfrac{(1-S_{11})(1-S_{22})-S_{12}S_{21}}{2S_{21}}$ $D = \dfrac{(1-S_{11})(1+S_{22})+S_{12}S_{21}}{2S_{21}}$	$\begin{bmatrix} \dfrac{Z_{11}}{Z_{21}} & \dfrac{\Delta^Z}{Z_{21}} \\[2mm] \dfrac{1}{Z_{21}} & \dfrac{Z_{22}}{Z_{21}} \end{bmatrix}$	$\begin{bmatrix} \dfrac{-Y_{22}}{Y_{21}} & \dfrac{-1}{Y_{21}} \\[2mm] \dfrac{-\Delta^Y}{Y_{21}} & \dfrac{-Y_{11}}{Y_{21}} \end{bmatrix}$	$\begin{bmatrix} \dfrac{-\Delta^h}{h_{21}} & \dfrac{-h_{11}}{h_{21}} \\[2mm] \dfrac{-h_{22}}{h_{21}} & \dfrac{-1}{h_{21}} \end{bmatrix}$	$\begin{bmatrix} V_1 \\ I_1 \end{bmatrix} = \begin{bmatrix} A & B \\ C & D \end{bmatrix}\begin{bmatrix} V_2 \\ -I_2 \end{bmatrix}$

three-terminal transistor can also be characterized by the three-port indefinite admittance matrix or indefinite scattering matrix given by

$$I_1 = y_{11}V_1 + y_{12}V_2 + y_{13}V_3 \qquad (1.57)$$

$$I_2 = y_{21}V_1 + y_{22}V_2 + y_{23}V_3 \qquad (1.58)$$

$$I_3 = y_{31}V_1 + y_{32}V_2 + y_{33}V_3 \qquad (1.59)$$

$$b_1 = S_{11}a_1 + S_{12}a_2 + S_{13}a_3 \qquad (1.60)$$

$$b_2 = S_{21}a_1 + S_{22}a_2 + S_{23}a_3 \qquad (1.61)$$

$$b_3 = S_{31}a_1 + S_{32}a_2 + S_{33}a_3 \qquad (1.62)$$

Since the reference terminal of a two-port is arbitrary, the three-port description is a more complete description of the transistor. However if $V_3 = 0$, the three-port reduces to a two-port, since $I_1 + I_2 + I_3 = 0$.

A well-known property of the indefinite admittance matrix is that the sum of any row or column is zero. Thus, if any set of four y-parameters is known, the remainder of the matrix is known. This allows one to pick a different reference node to calculate the y-parameters of a different two-port configuration. For example, if common-emitter y-parameters are measured, common-base y-parameters may be calculated.

In a similar manner, the indefinite scattering parameter matrix has the useful property that the sum of any row or column is unity. Thus measurement of common-emitter two-port S-parameters will completely characterize the transistor, since the indefinite scattering parameter matrix may be calculated from this data.

1.3   Conjugate Power Match

Consider a generator described by Γ_G and a load described by Γ_L connected to a transmission line of Z_0 characteristic impedance (Fig. 1.7). For maximum power transfer one must satisfy both

$$Z_L = Z_G^* \qquad (1.63)$$

and

$$Z_G = Z_0 \qquad (1.64)$$

Equation (1.63) is well-known conjugate impedance matching, and (1.64) expresses the requirement for the generator to deliver all of its available

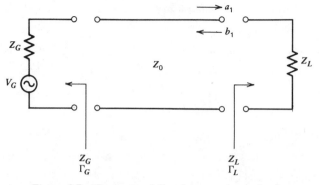

Figure 1.7 Generator delivering power to a load.

power to the transmission line. If (1.63) is satisfied, then

$$\Gamma_L = \Gamma_G^* \qquad (1.65)$$

is an equivalent way of expressing maximum power transfer. Alternatively, if a conjugate power match exists, then at any point on a transmission line the generator and load reflection coefficients are conjugate. Using the result of (1.11), we see that the conjugate power match for a lossless line can be expressed by

$$\Gamma_G = \left[\Gamma_L e^{-j2\beta l} \right]^* \qquad (1.66)$$

The mismatch loss is an important concept in the design of amplifiers. Normally the generator is matched to the transmission line, so that $\Gamma_G = 0$. In this case the conjugate mismatch loss is defined by the ratio of available power from the generator to the power delivered to the load:

$$M_c = \frac{P_A}{P_L} = \frac{P_A}{P_A - P_{\text{ref}}} = \frac{1}{1 - |\Gamma_L|^2} \geq 1 \qquad (1.67)$$

For the more general case of a reflecting generator and a reflecting load, we must know the available power from the generator. If the generator were connected to a nonreflecting load, the amplitude of the wave emitting from the generator would be b_G. In the next section we show that

$$a_1 = b_G + b_1 \Gamma_G \qquad (1.68)$$

The net power delivered to the load is

$$P_L = |a_1|^2 - |b_1|^2 = |a_1|^2 \left(1 - |\Gamma_L|^2\right) \tag{1.69}$$

Since $b_1 = a_1 \Gamma_L$ in Fig. 1.7, (1.68) gives

$$a_1 = b_G + a_1 \Gamma_L \Gamma_G \tag{1.70}$$

$$a_1 = \frac{b_G}{1 - \Gamma_L \Gamma_G} \tag{1.71}$$

Thus the power to the load may be written

$$P_L = \frac{|b_G|^2 \left(1 - |\Gamma_L|^2\right)}{|1 - \Gamma_L \Gamma_G|^2} \tag{1.72}$$

which is a function of b_G, Γ_L, and Γ_G. For a maximum power transfer (1.65) is satisfied, which is now substituted in (1.72) to give the available power from the generator:

$$P_A = \frac{|b_G|^2 \left(1 - |\Gamma_G^*|^2\right)}{|1 - \Gamma_G^* \Gamma_G|^2} = \frac{|b_G|^2}{1 - |\Gamma_G|^2} \tag{1.73}$$

Thus the conjugate mismatch loss in the general case is found from (1.72) and (1.73):

$$M_c = \frac{P_A}{P_L} = \frac{|1 - \Gamma_L \Gamma_G|^2}{\left(1 - |\Gamma_L|^2\right)\left(1 - |\Gamma_G|^2\right)} \geq 1 \tag{1.74}$$

The power delivered to the load may be calculated for this expression if Γ_L, Γ_G, and P_A are known.

1.4   Representation of the Generator

When a generator or source of power is connected to the two-port, the generator emits a wave b_G if a nonreflecting load is connected ($\Gamma_1 = 0$). In the general case where the load is not matched, consider Fig. 1.8. The first wave incident on the two-port is b_G, which is reflected as $b_G \Gamma_1$, which is reflected as $b_G \Gamma_1 \Gamma_G$, and so on. Thus the sum of the reflected waves coming

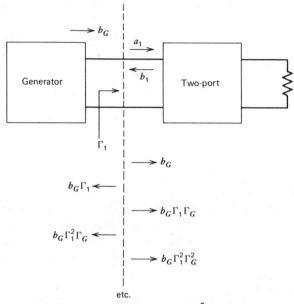

Figure 1.8 Wave reflections at generator port.

toward the generator is

$$b_1 = b_G \Gamma_1 \left[1 + \Gamma_1 \Gamma_G + (\Gamma_1 \Gamma_G)^2 + \cdots \right]$$

$$= \frac{b_G \Gamma_1}{1 - \Gamma_1 \Gamma_G} \tag{1.75}$$

Since $\Gamma_1 = b_1 / a_1$, (1.75) becomes

$$b_1 = \frac{b_G b_1}{a_1 - \Gamma_G b_1} \tag{1.76}$$

or

$$a_1 = b_G + b_1 \Gamma_G \tag{1.77}$$

This is an important relationship in S-parameter analysis. The incident wave on the two-port a_1 is not equal to b_G unless the load is nonreflecting, which would give $b_1 = 0$. An alternative view of the available power from the generator (1.73) will be derived using (1.77). Since $\Gamma_1 = b_1/a_1$, (1.77) gives

$$a_1 = b_G + \Gamma_1 \Gamma_G a_1 \tag{1.78}$$

$$a_1 = \frac{b_G}{1 - \Gamma_1 \Gamma_G} \tag{1.79}$$

For the available power to be the incident power, the equation $\Gamma_1 = \Gamma_G^*$ may be used in (1.79) to give

$$P_A = |a_1|^2 \left(1 - |\Gamma_G|^2\right) = \frac{|b_G|^2}{1 - |\Gamma_G|^2} \tag{1.80}$$

which is the same result given in (1.73). For a nonreflecting load ($\Gamma_L = \Gamma_1 = 0$), (1.72) gives $P_L = |b_G|^2$, which is always less than the available power given by (1.80). Another equivalent representation of the generator is given in Fig. 1.9. The voltage on the transmission line Z_0 can be related to the generator voltage V_G. For the conjugately matched case in Fig. 1.9, the reflected voltage is zero ($V_{ref} = 0$). Thus the waves on the line reduce to

$$V_{inc} e^{-\gamma x}$$

$$I_{inc} e^{-\gamma x} = \frac{V_{inc}}{Z_0} e^{-\gamma x}$$

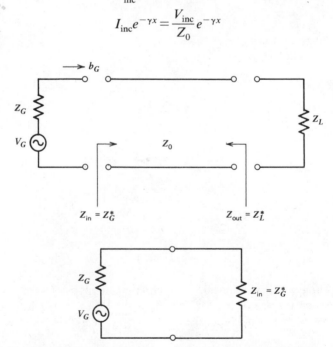

Figure 1.9   Conjugate matched generator and load for maximum power transfer.

In this case the incident power is the available power from the generator and is given by

$$P_{\text{inc}} = P_A = V_{\text{inc}} I^*_{\text{inc}} = \frac{V^2_{\text{inc}}}{Z_0} \tag{1.81}$$

Since

$$P_A = \frac{V^2_G}{4\,\text{Re}(Z_G)} \tag{1.82}$$

We have

$$V_{\text{inc}} = \frac{V_G}{2} \tag{1.83}$$

for the conjugately matched case.

Referring to (1.29)–(1.32), the definitions of a, b, and so on, we have

$$b_G = \frac{V_{\text{inc}}}{\sqrt{Z_0}} = \frac{V_G}{2\sqrt{Z_0}} \tag{1.84}$$

where the factor 2 has come from maximum power transfer considerations.

In the more general case where a conjugate match does not exist, we find that b_G is given by

$$b_G = \frac{V_G}{\sqrt{Z_0}}\frac{Z_0}{Z_G + Z_0} = \frac{V_G\sqrt{Z_0}}{Z_G + Z_0} \tag{1.85}$$

A proof that (1.85) is true follows. Substituting (1.85) into (1.73) and using (1.7) for Γ_G gives

$$P_A = \frac{V^2_G Z_0}{|Z_G + Z_0|^2 (1 - |\Gamma_G|^2)}$$

$$= \frac{V^2_G Z_0}{(Z_G + Z_0)(Z^*_G + Z_0)\left\{1 - \dfrac{[(Z_G - Z_0)(Z^*_G - Z_0)]}{[(Z_G + Z_0)(Z^*_G + Z_0)]}\right\}}$$

$$= \frac{V^2_G}{4\,\text{Re}(Z_G)}$$

which is the same as (1.82).

The source or generator wave b_G can be represented either by (1.77) in terms of a_1, b_1, and Γ_G or by (1.85) in terms of V_G, Z_G, and Z_0. The available power from the generator is given by either (1.73) or (1.82).

1.5 Stability

The amplifier design problem (Fig. 1.1) is very simple when the two-port transistor is unconditionally stable. In this case a simultaneous conjugate match at the input and output ports will deliver the maximum power to the load and therefore will produce the highest power gain.

The question of stability can be considered from three points of view:

1 In the Γ_L plane, what values of Γ_L give $|S_{11}'| > 1$?
2 In the S_{11}' plane, where does $|\Gamma_L| = 1$ plot?
3 If $(S_{11}')^* = \Gamma_G$ and $(S_{22}')^* = \Gamma_L$, the resistors terminating the network are positive.

All of these points of view will lead to the same conclusion concerning the stability of the two-port. If both conditions (1) and (2) give unconditional stability, the necessary and sufficient conditions are satisfied for unconditional stability. The third condition will be discussed in Section 1.6.

The conditions for two-port stability are

$$|S_{11}'| < 1 \tag{1.86}$$

$$|S_{22}'| < 1 \tag{1.87}$$

for all possible load terminations with a positive real part (i.e., positive resistors). If either of the conditions in (1.86) and (1.87) can be violated, the two-port is only conditionally stable. The condition $|S_{11}'| > 1$ or $|S_{22}'| > 1$ is equivalent to a negative resistance at that port, which is needed for oscillator design.

From previous considerations [(1.36) and (1.38)] the values of S_{11}' and S_{22}' are

$$S_{11}' = S_{11} + \frac{S_{12}S_{21}\Gamma_L}{1 - S_{22}\Gamma_L} = \frac{S_{11} - D\Gamma_L}{1 - S_{22}\Gamma_L} \tag{1.88}$$

$$S_{22}' = S_{22} + \frac{S_{12}S_{21}\Gamma_G}{1 - S_{11}\Gamma_G} = \frac{S_{22} - D\Gamma_G}{1 - S_{11}\Gamma_G} \tag{1.89}$$

$$D = S_{11}S_{22} - S_{12}S_{21} \tag{1.90}$$

For unconditional stability, all values of Γ_L must insure that the input reflection coefficient S_{11}' is less than unity and all values of Γ_G must insure that the output reflection coefficient S_{22}' is less than unity.

Consider the mapping of $|S_{11}'| = 1$ in the Γ_L plane, as shown in Fig. 1.10. Using the following substitutions in (1.88) will give the conditions for stability:

$$S_{11} = S_{11R} + jS_{11I} \tag{1.91}$$

$$S_{22} = S_{22R} + jS_{22I} \tag{1.92}$$

$$D = D_R + jD_I \tag{1.93}$$

$$\Gamma_L = U_2 + jV_2 \tag{1.94}$$

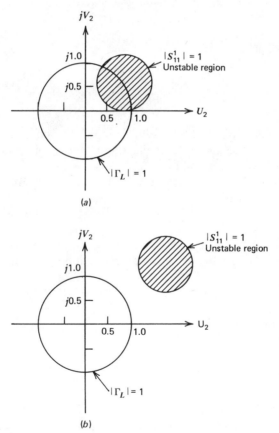

Figure 1.10 Stability viewed in the Γ_L plane. (*a*) Conditionally stable; (*b*) unconditionally stable.

From (1.86) and (1.88), the boundary for stability is

$$|S_{11} - D\Gamma_L| < |1 - S_{22}\Gamma_L| \tag{1.95}$$

Substituting (1.91)–(1.94) and squaring both sides gives the following circle:

$$(U_2 - U_{2G})^2 + (V_2 - V_{2G})^2 = r_G^2 \tag{1.96}$$

where the center is

$$
\begin{aligned}
C_G &= U_{2G} + jV_{2G} \\
&= \frac{(S_{22} - DS_{11}^*)^*}{|S_{22}|^2 - |D|^2}
\end{aligned}
\tag{1.97}
$$

and the radius is

$$r_G = \frac{|S_{12}S_{21}|}{||S_{22}|^2 - |D|^2|} \tag{1.98}$$

The inside of this circle is the unstable region if the point $U_2 + jV_2 = 0$ gives stability, that is, if $|S_{11}'| < 1$. A similar result can be found in the Γ_G plane for $S_{22}' = 1$ by using (1.87) and (1.89).

For an unconditionally stable two-port, the geometry of Fig. 1.10(b) is satisfied:

$$|C_G| - r_G > 1 \tag{1.99}$$

which finally produces the result (a derivation appears in Appendix A):

$$1 - |S_{11}|^2 - |S_{22}|^2 + |D|^2 > 2|S_{12}||S_{21}| \tag{1.100}$$

If we define the stability factor k as

$$k = \frac{1 - |S_{11}|^2 - |S_{22}|^2 + |D|^2}{2|S_{12}||S_{21}|} \tag{1.101}$$

a stability factor greater than unity is required for unconditional stability.

For the second view of unconditional stability, plot the circle $\Gamma_L = 1$ in S_{11}' plane, as shown in Fig. 1.11. This again produces a circle with radius

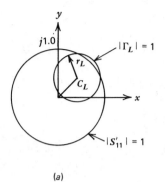

(a)

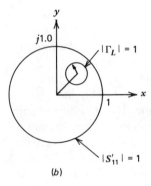

(b)

Figure 1.11 Stability viewed in the S_{11} plane. (a) Conditionally stable; (b) unconditionally stable.

and center given by

$$C_L = S_{11} + \frac{S_{12}S_{21}S_{22}^*}{1 - |S_{22}|^2} \qquad (1.102)$$

$$r_L = \frac{|S_{12}S_{21}|}{1 - |S_{22}|^2} \qquad (1.103)$$

Since C_L could be zero,

$$r_L < 1 \qquad (1.104)$$

is required for stability.

$$|S_{12}S_{21}| < 1 - |S_{22}|^2 \qquad (1.105)$$

and similarly for the $\Gamma_G = 1$ circle in the S'_{22} plane:

$$|S_{12}S_{21}| < 1 - |S_{11}|^2 \qquad (1.106)$$

In summary, the conditions (necessary and sufficient) for unconditional stability are

$$k = \frac{1 - |S_{11}|^2 - |S_{22}|^2 + |D|^2}{2|S_{12}||S_{21}|} > 1 \qquad (1.107)$$

$$|S_{12}S_{21}| < 1 - |S_{11}|^2 \qquad (1.108)$$

$$|S_{12}S_{21}| < 1 - |S_{22}|^2 \qquad (1.109)$$

Since normally $|D| < 1$, one can then show that $k > 1$ is sufficient to guarantee unconditional stability. Otherwise, three conditions are required (necessary and sufficient) for absolute unconditional stability.

The first step in amplifier or oscillator design is to determine the stability factor versus frequency. For the regions where $k < 1$, the stability circles are plotted in the Γ_L plane and in the Γ_G plane. The matching circuits in Fig. 1.1 must be designed to avoid the unstable regions given by stability circles. Normally, the low-frequency range of the transistor will give $k < 1$, so the low-frequency value of Γ_L and Γ_G must be carefully chosen. A short circuit termination at low frequencies will usually guarantee low-frequency stability.

1.6   Simultaneous Conjugate Match

For a two-port with the stability factor greater than unity, it is possible to simultaneously conjugately match the two-port to produce the maximum possible gain (G_{ma}). The conditions for simultaneous conjugate match are

$$S'_{11} = S_{11} + \frac{S_{12}S_{21}\Gamma_L}{1 - S_{22}\Gamma_L} = \Gamma_G^* \qquad (1.110)$$

and

$$S'_{22} = S_{22} + \frac{S_{12}S_{21}\Gamma_G}{1 - S_{11}\Gamma_G} = \Gamma_L^* \qquad (1.111)$$

which become

$$(1 - \Gamma_L S_{22})(S_{11} - \Gamma_G^*) + \Gamma_L S_{12}S_{21} = 0 \qquad (1.112)$$

and

$$(1 - \Gamma_G S_{11})(S_{22} - \Gamma_L^*) + \Gamma_G S_{12} S_{21} = 0 \qquad (1.113)$$

From (1.112) and (1.113) we find

$$\Gamma_L = \frac{\Gamma_G^* - S_{11}}{\Gamma_G^* S_{22} - D} \qquad (1.114)$$

$$\Gamma_G = \frac{\Gamma_L^* - S_{22}}{\Gamma_L^* S_{11} - D} \qquad (1.115)$$

Substituting in (1.113) gives

$$\Gamma_G^2 - \Gamma_G \frac{B_1}{C_1} + \frac{C_1^*}{C_1} = 0 \qquad (1.116)$$

$$C_1 = S_{11} - DS_{22}^* \qquad (1.117)$$

$$B_1 = 1 - |S_{22}|^2 + |S_{11}|^2 - |D|^2 \qquad (1.118)$$

The solution for (1.116) is

$$\begin{aligned}
\Gamma_{Gm} &= \frac{B_1}{2C_1} \pm \frac{1}{2} \sqrt{\left(\frac{B_1}{C_1}\right)^2 - 4\frac{C_1^*}{C_1}} \\
&= \frac{C_1^*}{|C_1|} \left[\frac{B_1}{2|C_1|} \pm \sqrt{\frac{B_1^2}{|2C_1|^2} - 1} \right]
\end{aligned} \qquad (1.119)$$

In examining (1.119) there are four cases to consider, as shown in Table 1.2. Since it can be shown that $B_1/2|C_1| > 1$ corresponds to $k > 1$, this is the only case that can produce a useful solution. This statement is proved in Appendix B.

Also, $B_1 < 0$ can only occur if the two-port is potentially unstable, that is, if (1.107)–(1.109) are not satisfied. Thus the only useful solution from Table 1.2 is Case 1, which gives

$$\Gamma_{Gm} = \frac{C_1^*}{|C_1|} \left[\frac{B_1}{2|C_1|} - \sqrt{\frac{B_1^2}{|2C_1|^2} - 1} \right] \qquad (1.120)$$

Table 1.2   Four Cases of (1.119)

$B_1 > 0$ Normal Condition		$B_1 < 0$	
Case 1	Case 2	Case 3	Case 4
$\dfrac{B_1}{2|C_1|} > 1$	$\dfrac{B_1}{2|C_1|} < 1$	$\dfrac{B_1}{2|C_1|} > 1$	$\dfrac{B_1}{2|C_1|} < 1$
$k > 1$	$k < 1$	$k > 1$	$k < 1$
Useful solution given in (1.120)	$|\Gamma_{Gm}| = 1$ not useful	Potentially unstable even though $k > 1$	$|\Gamma_{Gm}| = 1$ not useful

Notice that for very small S_{12} and small D, $C_1 \simeq S_{11}$.

$$\left.\Gamma_{Gm}\right|_{S_{12}=0} \simeq \frac{S_{11}^*}{|C_1|}\left[\frac{B_1}{2|C_1|} - \sqrt{\frac{B_1^2}{|2C_1|^2} - 1}\right] \qquad (1.121)$$

which shows that the optimum value of Γ_{Gm} may be near S_{11}^*. Note the angle is essentially the angle of S_{11}^*. A similar result for the load termination is

$$\Gamma_{Lm} = \frac{C_2^*}{|C_2|}\left(\frac{B_2}{2|C_2|} - \sqrt{\frac{B_2^2}{|2C_2|^2} - 1}\right) \qquad (1.222)$$

$$C_2 = S_{22} - DS_{11}^* \qquad (1.123)$$

$$B_2 = 1 - |S_{11}|^2 + |S_{22}|^2 - |D|^2 \qquad (1.124)$$

$$\left.\Gamma_{Lm}\right|_{S_{12}=0} \simeq \frac{S_{22}^*}{|C_2|}\left[\frac{B_2}{2|C_2|} - \sqrt{\frac{B_2^2}{|2C_2|^2} - 1}\right] \qquad (1.125)$$

Useful solutions are possible only when

$$\frac{B_1}{|2C_1|} > 1 \qquad (1.126)$$

$$\frac{B_2}{|2C_2|} > 1 \qquad (1.127)$$

which implies that $k > 1$. A proof appears in Appendix B.

1.7 Power Gains

Several definitions of power gains are important in the design of amplifiers. Consider the power flow from generator to load shown in Fig. 1.12, where M_1 and M_2 are lossless matching networks. Three power gains may be defined. The transducer power gain is the most significant, since it shows the insertion effect of the total amplifier. If the amplifier is omitted, $P_A = P_L$ (for a conjugate generator and load impedance). The transducer power gain is defined by

$$G_T = G_T(\Gamma_G, \Gamma_L, S) = \frac{P_L}{P_A} \qquad (1.128)$$

where S is the S-parameter matrix of the two-port. The available power gain is defined by

$$G_A = G_A(\Gamma_G, S) = \frac{P_{avo}}{P_A} \qquad (1.129)$$

which is not a function of the load match Γ_L. The power gain is defined by

$$G = G(\Gamma_L, S) = \frac{P_L}{P_{in}} \qquad (1.130)$$

which is not a function of the generator match Γ_G. The transducer gain can only approach the available power gain and the power gain:

$\qquad G_T \leq G_A \qquad$ Where equality occurs for a Γ_L conjugate match

$\qquad G_T \leq G \qquad$ Where equality occurs for a Γ_G conjugate match

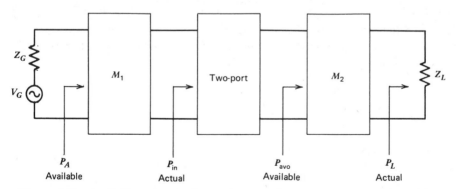

Figure 1.12 Available power and actual power for a two-port connected to a generator and a load.

For a simultaneous conjugate match at both ports, the transducer power gain is the maximum available gain G_{ma} given by (1.161). This is possible only for $k > 1$.

The transducer power gain is derived below. Using an Eqn. similar to (1.69) and (1.73) in (1.128) gives

$$G_T(\Gamma_G, \Gamma_L, S) = \frac{|b_2|^2}{|b_G|^2}\left(1 - |\Gamma_L|^2\right)\left(1 - |\Gamma_G|^2\right) \tag{1.131}$$

Using (1.33), (1.34), and (1.79) gives

$$\frac{b_2}{a_1} = \frac{S_{21}}{1 - S_{22}\Gamma_L} \tag{1.132}$$

$$\frac{a_1}{b_G} = \frac{1}{1 - \Gamma_1\Gamma_G} \tag{1.133}$$

$$G_T = \frac{\left(1 - |\Gamma_L|^2\right)|S_{21}|^2\left(1 - |\Gamma_G|^2\right)}{|1 - S_{22}\Gamma_L|^2|1 - \Gamma_1\Gamma_G|^2} \tag{1.134}$$

$$\Gamma_1 = S'_{11} = S_{11} + \frac{S_{12}S_{21}\Gamma_L}{1 - S_{22}\Gamma_L} \tag{1.135}$$

$$G_T = \frac{\left(1 - |\Gamma_L|^2\right)|S_{21}|^2\left(1 - |\Gamma_G|^2\right)}{|1 - S_{22}\Gamma_L|^2|1 - S_{11}\Gamma_G - (S_{12}S_{21}\Gamma_L\Gamma_G)/(1 - S_{22}\Gamma_L)|^2}$$

$$= \frac{\left(1 - |\Gamma_L|^2\right)|S_{21}|^2\left(1 - |\Gamma_G|^2\right)}{|(1 - S_{22}\Gamma_L)(1 - S_{11}\Gamma_G) - S_{12}S_{21}\Gamma_L\Gamma_G|^2} \tag{1.136}$$

This is the exact solution for transducer power gain as a function of Γ_G, Γ_L, and all four S-parameters. Comparing (1.134) and (1.136), we can also express G_T as

$$G_T = \frac{\left(1 - |\Gamma_L|^2\right)|S_{21}|^2\left(1 - |\Gamma_G|^2\right)}{|1 - \Gamma_2\Gamma_L|^2|1 - S_{11}\Gamma_G|^2} \tag{1.137}$$

$$\Gamma_2 = S'_{22} = S_{22} + \frac{S_{12}S_{21}\Gamma_G}{1 - S_{11}\Gamma_G} \tag{1.138}$$

The available power gain follows by substituting $\Gamma_L = \Gamma_2^*$ in (1.137). This gives

$$G_A = G_A(\Gamma_G, S) = \frac{\left(1 - |\Gamma_G|^2\right)|S_{21}|^2}{|1 - S_{11}\Gamma_G|^2\left(1 - |S_{22}'|^2\right)} \tag{1.139}$$

which is a function of Γ_G and all four S-parameters. The power gain follows by substituting $\Gamma_G = \Gamma_1^*$ in (1.134). This gives

$$G = G(\Gamma_L, S) = \frac{\left(1 - |\Gamma_L|^2\right)|S_{21}|^2}{|1 - S_{22}\Gamma_L|^2\left(1 - |S_{11}'|^2\right)} \tag{1.140}$$

which is a function of Γ_L and all four S-parameters.

A useful approximation for transducer power gain is to assume that $S_{12} = 0$ in (1.136) to give G_{TU}, the unilateral transducer power gain:

$$G_{TU} = \frac{\left(1 - |\Gamma_L|^2\right)|S_{21}|^2\left(1 - |\Gamma_G|^2\right)}{|1 - S_{22}\Gamma_L|^2|1 - S_{11}\Gamma_G|^2} \tag{1.141}$$

The effect of Γ_L mismatch and Γ_G mismatch can be clearly seen in this equation. For an input and output match, $\Gamma_L = S_{22}^*$ and $\Gamma_G = S_{11}^*$ this reduces to

$$G_{TUmax} = \frac{|S_{21}|^2}{\left(1 - |S_{22}|^2\right)\left(1 - |S_{11}|^2\right)} \tag{1.142}$$

which is a useful approximation for the maximum gain with simultaneous conjugate match. This form is simple to calculate and shows the importance of input and output mismatch, which is inherent at the input and output ports. If the input transistor mismatch is

$$T_1 = \frac{1}{1 - |S_{11}|^2} \tag{1.143}$$

and the output transistor mismatch is

$$T_2 = \frac{1}{1 - |S_{22}|^2} \tag{1.144}$$

the maximum unilateral transducer power gain is

$$G_{TUmax} = T_1|S_{21}|^2 T_2 \tag{1.145}$$

For a two-port that is conditionally stable ($k < 1$), the maximum achievable gain is G_{ms} given by

$$G_{ms} = \frac{|S_{21}|}{|S_{12}|} = \frac{|y_{21}|}{|y_{12}|} = \cdots \tag{1.146}$$

This is the gain that can be achieved by resistively loading the two-port such that $k = 1$ and then simultaneously conjugately matching the input and output ports. For conditionally stable two-ports, the maximum stable gain is an upper limit in power gain that can only be approached as the input and output mismatch is reduced. If a simultaneous conjugate match is attempted, the two-port will oscillate if $k < 1$.

The maximum available gain can be derived by considering the power gain as a function of the Γ_L termination. If the two-port is unconditionally stable, this power gain can be maximized by proper choice of the load termination. Using (1.140) for power gain and expressing S'_{11} as follows

$$S'_{11} = S_{11} + \frac{S_{12}S_{21}\Gamma_L}{1 - S_{22}\Gamma_L} = \frac{S_{11} - \Gamma_L D}{1 - S_{22}\Gamma_L} \tag{1.147}$$

gives

$$G(\Gamma_L, S) = \frac{|S_{21}|^2 \left(1 - |\Gamma_L|^2\right)}{|1 - S_{22}\Gamma_L|^2 - |S_{11} - \Gamma_L D|^2} \tag{1.148}$$

Expanding the denominator gives

$$G(\Gamma_L, S) = \frac{|S_{21}|^2 \left(1 - |\Gamma_L|^2\right)}{\left(1 - |S_{11}|^2\right) + |\Gamma_L|^2 \left(|S_{22}|^2 - |D|^2\right) - 2\mathrm{Re}(\Gamma_L C_2)} \tag{1.149}$$

$$C_2 = S_{22} - DS_{11}^* \tag{1.150}$$

The power gain is a function of Γ_L and is expressed as

$$G(\Gamma_L, S) = |S_{21}|^2 g_2 \tag{1.151}$$

$$\Gamma_L = U_2 + jV_2 \tag{1.152}$$

$$\mathrm{Re}(\Gamma_L C_2) = \mathrm{Re}[U_2 + jV_2][\mathrm{Re}\, C_2 + j\,\mathrm{Im}\, C_2]$$

$$= U_2 \mathrm{Re}\, C_2 - V_2 \mathrm{Im}\, C_2 \tag{1.153}$$

Therefore

$$g_2 = \frac{|1 - U_2^2 - V_2^2|}{\left|1 - |S_{11}|^2 + (U_2^2 + V_2^2)(|S_{22}|^2 - |D|^2) - 2U_2 \operatorname{Re} C_2 + 2V_2 \operatorname{Im} C_2\right|}$$

(1.154)

This can be expressed as a circle in the Γ_L plane in the form

$$(U_2 - U_{2c})^2 + (V_2 - V_{2c})^2 = \rho_{2c}^2$$

(1.155)

Expanding (1.154) gives

$$U_2^2 + V_2^2 + 2\frac{U_2 \operatorname{Re} C_2 g_2}{-1 - |S_{22}|g_2 + |D|g_2} - 2\frac{V_2 \operatorname{Im} C_2 g_2}{-1 - |S_{22}|g_2 + |D|g_2}$$

$$= \frac{-1 + g_2 - |S_{11}|^2 g_2}{-1 - |S_{22}|^2 g_2 + |D|^2 g_2}$$

(1.156)

Adding the terms

$$\frac{\operatorname{Re} C_2^2 g_2^2 + \operatorname{Im} C_2^2 g_2^2}{\left|-1 - |S_{22}|^2 g_2 + |D|^2 g_2\right|^2}$$

to both sides of (1.156) gives the radius of the circle, where the following identity is also used:

$$\operatorname{Re} C_2^2 + \operatorname{Im} C_2^2 = |S_{22} - DS_{11}^*|^2 = |S_{12} S_{21}|^2 + (1 - |S_{11}|^2)(|S_{22}|^2 - |D|^2)$$

(1.157)

Regrouping terms as in (1.155) gives

$$\rho_{2c}^2 = \left| \frac{-1 + g_2 - |S_{11}|^2 g_2}{-1 - |S_{22}|^2 g_2 + |D|^2 g_2} + \frac{|S_{22} - DS_{11}^*|^2}{\left|-1 - |S_{22}|^2 g_2 + |D|^2 g_2\right|^2} \right|$$

$$= \frac{|1 + g_2\{|S_{22}|^2 + |S_{11}|^2 - |D|^2 - 1\} + g_2^2\{-|S_{22}|^2 + |D|^2}{+ |S_{11}|^2 |S_{22}|^2 - |S_{11}|^2 |D|^2 + |S_{12} S_{21}|^2 + |S_{22}|^2 - |D|^2}{\frac{- |S_{11}|^2 |S_{22}|^2 + |S_{11}|^2 |D|^2\}|}{\left|-1 - |S_{22}|^2 g_2 + |D|^2 g_2\right|^2}$$

$$= \frac{1 - 2k|S_{12} S_{21}|g_2 + |S_{12} S_{21}|^2 g_2^2}{\left|-1 - |S_{22}|^2 g_2 + |D|^2 g_2\right|^2}$$

(1.158)

To maximize the gain, set $\rho_{2C}=0$ and solve for g_{20}:

$$g_{20}^2 - \frac{2k}{|S_{12}S_{21}|}g_{20} + \frac{1}{|S_{12}S_{21}|^2} = 0 \qquad (1.159)$$

$$g_{20} = \frac{1}{|S_{12}S_{21}|}\left(k \pm \sqrt{k^2-1}\right) \qquad (1.160)$$

Thus the maximum available gain is

$$G_{\mathrm{ma}} = \left|\frac{S_{21}}{S_{12}}\right|\left(k \pm \sqrt{k^2-1}\right) \qquad (1.161)$$

where $k>1$. The sign in (1.161) is usually negative, since B_1 is usually positive. B_1 was given by (1.119). Although G_{ma} is called the "maximum available gain," it is also the maximum power gain of (1.130) and the maximum transducer power gain of (1.136).

One additional power gain is the unilateral power gain. This is the maximum available power gain when the two-port has been simultaneously conjugately matched and the feedback parameter has been neutralized to zero. The conditions are indicated in Fig. 1.13 for unilateral power gain. Notice that for $S_{12}=0$, the stability factor is infinite, but G_{ma} does exist [see (1.142)]. From the general considerations of passivity of a two-port, the network must have $U>1$ for net power flowing out of the two-port. This

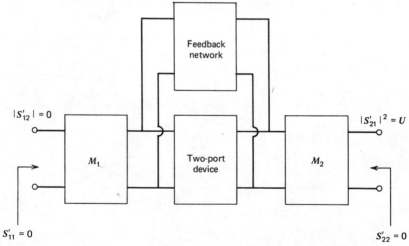

Figure 1.13 Conditions for unilateral power gain.

may be expressed in y- or z-parameters of the two-port device as

$$U = \frac{|z_{21} - z_{12}|^2}{4\,\mathrm{Re}\,z_{11}\,\mathrm{Re}\,z_{22} - 4\,\mathrm{Re}\,z_{12}z_{21}} \tag{1.162}$$

Since the stability factor k is also expressed as

$$k = \frac{2\,\mathrm{Re}\,z_{11}\,\mathrm{Re}\,z_{22} - \mathrm{Re}(z_{12}z_{21})}{|z_{12}z_{21}|} \tag{1.163}$$

the unilateral gain may be expressed as

$$U = \frac{|z_{21} - z_{12}|^2}{2k|z_{12}z_{21}| - 2\,\mathrm{Re}\,z_{12}\,\mathrm{Re}\,z_{21} - 2\,\mathrm{Im}\,z_{12}\,\mathrm{Im}\,z_{21}}$$

$$= \frac{|z_{21}/z_{12} - 1|^2}{2k|z_{21}/z_{12}| - 2\,\mathrm{Re}(z_{12}^* z_{21})/|z_{12}|^2}$$

$$= \frac{|z_{21}/z_{12} - 1|^2}{2k|z_{21}/z_{12}| - 2\,\mathrm{Re}|z_{21}/z_{12}|} \tag{1.164}$$

Now using the conversion formula given in Table 1.1, we find that the unilateral gain is

$$U = \frac{|S_{21}/S_{12} - 1|^2}{2k|S_{21}/S_{12}| - 2\,\mathrm{Re}|S_{21}/S_{12}|} \tag{1.165}$$

This gain is the highest possible gain that the active two-port could ever achieve. The frequency where the unilateral gain becomes unity defines the boundary between an active and a passive network. This frequency is usually referred to as $f_{\max}$, the maximum frequency of oscillation. If S_{12} of the two-port device is set equal to zero in (1.165), one may again derive G_{TUmax} given by (1.142).

$$U|_{S_{12}=0} = \frac{|S_{21}/S_{12}||S_{21}/S_{12}|}{2k|S_{21}/S_{12}|}$$

$$= \frac{|S_{21}/S_{12}||S_{21}/S_{12}|}{\{[1 - |S_{11}|^2][1 - |S_{22}|^2]\}/(|S_{12}||S_{21}|)|S_{21}/S_{12}|}$$

$$= \frac{|S_{21}|^2}{[1 - |S_{11}|^2][1 - |S_{22}|^2]} = G_{\mathrm{TUmax}} \tag{1.166}$$

Table 1.3 Nine Power Gains

Transducer power gain in 50-ohm system	$G_T = \lvert S_{21} \rvert^2$
Transducer power gain for arbitrary Γ_G and Γ_L	$G_T = \dfrac{\left(1 - \lvert \Gamma_G \rvert^2\right) \lvert S_{21} \rvert^2 \left(1 - \lvert \Gamma_L \rvert^2\right)}{\lvert (1 - S_{11}\Gamma_G)(1 - S_{22}\Gamma_L) - S_{12}S_{21}\Gamma_G\Gamma_L \rvert^2}$
Unilateral transducer power gain	$G_{\text{TU}} = \dfrac{\lvert S_{21} \rvert^2 \left(1 - \lvert \Gamma_G \rvert^2\right)\left(1 - \lvert \Gamma_L \rvert^2\right)}{\lvert 1 - S_{11}\Gamma_G \rvert^2 \lvert 1 - S_{22}\Gamma_L \rvert^2}$
Power gain with input conjugate matched	$G = \dfrac{\lvert S_{21} \rvert^2 \left(1 - \lvert \Gamma_L \rvert^2\right)}{\lvert 1 - S_{22}\Gamma_L \rvert \left(1 - \lvert S'_{11} \rvert^2\right)} = \dfrac{\lvert S_{21} \rvert^2}{1 - \lvert S_{11} \rvert^2}$ (for $\Gamma_L = 0$)
Available power gain with output conjugate matched	$G_A = \dfrac{\lvert S_{21} \rvert^2 \left(1 - \lvert \Gamma_G \rvert^2\right)}{\lvert 1 - S_{11}\Gamma_G \rvert^2 \left(1 - \lvert S'_{22} \rvert^2\right)} = \dfrac{\lvert S_{21} \rvert^2}{1 - \lvert S_{22} \rvert^2}$ (for $\Gamma_G = 0$)
Maximum available power gain	$G_{\text{ma}} = \left\lvert \dfrac{S_{21}}{S_{12}} \right\rvert \left(k - \sqrt{k^2 - 1}\right)$
Maximum unilateral transducer power gain	$G_{\text{TU max}} = \dfrac{\lvert S_{21} \rvert^2}{\left(1 - \lvert S_{11} \rvert^2\right)\left(1 - \lvert S_{22} \rvert^2\right)}$
Maximum stable power gain	$G_{\text{ms}} = \dfrac{\lvert S_{21} \rvert}{\lvert S_{12} \rvert}$
Unilateral power gain	$U = \dfrac{1/2 \lvert S_{21}/S_{12} - 1 \rvert^2}{k \lvert S_{21}/S_{12} \rvert - \operatorname{Re}(S_{21}/S_{12})}$

All of the power gains defined in this chapter can be expressed in decibels. A summary of the power gains appears in Table 1.3. In decibels,

$$U_{\text{dB}} = 10 \log U \tag{1.167}$$

However, for the case of a transducer gain given by S_{21}, this is a voltage

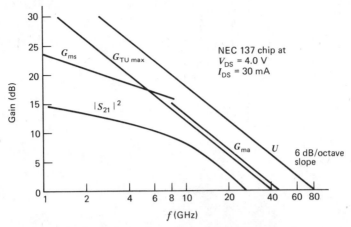

Figure 1.14 Power gains for a GaAs MESFET versus frequency.

gain, so it must be squared to give the transducer power gain:

$$G_{TdB} = |S_{21}|^2_{dB} = 10\log|S_{21}|^2$$

$$= 20\log|S_{21}| \tag{1.168}$$

The power gains for a typical common-source GaAs MESFET are plotted versus frequency in Figure 1.14. This figure indicates that the largest gain is always U and that G_{ma} will exist only for $k > 1$.

1.8 Three-Port S-Parameters

A three-port design may be required if a third port is available for tuning or passive termination. An example is the dual-gate GaAs MEtal Semiconductor Field-Effect Transistor (MESFET), which usually operates in the common-source mode with port 1 as gate 1, port 2 as drain, and port 3 as gate 2. The three-port is identified in Fig. 1.15.

The transistor must be described by a three-port set of S-parameters given by

$$b_1 = S_{11}a_1 + S_{12}a_2 + S_{13}a_3 \tag{1.169}$$

$$b_2 = S_{21}a_1 + S_{22}a_2 + S_{23}a_3 \tag{1.170}$$

$$b_3 = S_{31}a_1 + S_{32}a_2 + S_{33}a_3 \tag{1.171}$$

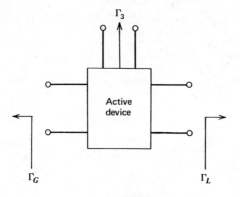

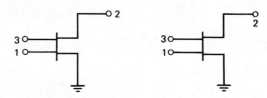

Figure 1.15   Three-port networks.

Since the termination at each port is given by

$$a_1 = b_1 \Gamma_G \tag{1.172}$$

$$a_2 = b_2 \Gamma_L \tag{1.173}$$

$$a_3 = b_3 \Gamma_3 \tag{1.174}$$

a straightforward calculation similar to (1.37) and (1.38) will give

$$S'_{11} = \frac{b_1}{a_1} = S_{11} + S_{12} \frac{S_{21}\Gamma_L + \Gamma_L\Gamma_3(S_{23}S_{31} - S_{21}S_{33})}{(1 - S_{22}\Gamma_L)(1 - S_{33}\Gamma_3) - S_{23}S_{32}\Gamma_3\Gamma_L}$$

$$+ S_{13} \frac{S_{31}\Gamma_3 + \Gamma_L\Gamma_3(S_{32}S_{21} - S_{31}S_{22})}{(1 - S_{22}\Gamma_L)(1 - S_{33}\Gamma_3) - S_{23}S_{32}\Gamma_3\Gamma_L} \tag{1.175}$$

$$S'_{22} = \frac{b_2}{a_2} = S_{22} + S_{21} \frac{S_{12}\Gamma_G + \Gamma_G\Gamma_3(S_{13}S_{32} - S_{33}S_{12})}{(1 - S_{11}\Gamma_G)(1 - S_{33}\Gamma_3) - S_{13}S_{31}\Gamma_3\Gamma_G}$$

$$+ S_{23} \frac{S_{32}\Gamma_3 + \Gamma_G\Gamma_3(S_{31}S_{12} - S_{11}S_{32})}{(1 - S_{11}\Gamma_G)(1 - S_{33}\Gamma_3) - S_{13}S_{31}\Gamma_3\Gamma_G} \tag{1.176}$$

$$S_{33}' = \frac{b_3}{a_3} = S_{33} + S_{31} \frac{S_{13}\Gamma_G + \Gamma_L\Gamma_G(S_{12}S_{23} - S_{22}S_{13})}{(1 - S_{11}\Gamma_G)(1 - S_{22}\Gamma_L) - S_{12}S_{21}\Gamma_L\Gamma_G}$$

$$+ S_{32} \frac{S_{23}\Gamma_L + \Gamma_L\Gamma_G(S_{21}S_{13} - S_{11}S_{23})}{(1 - S_{11}\Gamma_G)(1 - S_{22}\Gamma_L) - S_{12}S_{21}\Gamma_L\Gamma_G} \qquad (1.177)$$

Because of the mathematical complexity, the three-port is usually reduced to a two-port by terminating port three. For dual-gate amplifiers an rf short circuit at port three is often chosen for a two-port stability factor greater than unity.

$$\Gamma_3 = -1 \qquad (1.178)$$

The two-port S-parameters of this network can be derived from (1.169)–(1.171). Alternatively, an rf short could be connected to gate 2 and the two-port S-parameters of the transistor could be measured.

The effect of the third-port termination can be investigated if $\Gamma_L = 0$. This reduces (1.175) to

$$S_{11}' = S_{11} + \frac{S_{13}S_{31}\Gamma_3}{1 - S_{33}\Gamma_3} \qquad (1.179)$$

Likewise, if $\Gamma_G = 0$, (1.176) reduces to

$$S_{22}' = S_{22} + \frac{S_{23}S_{32}\Gamma_3}{1 - S_{33}\Gamma_3} \qquad (1.180)$$

In the more general case of arbitrary choice of Γ_G, Γ_L, and Γ_3, a large number of interesting amplifier and oscillator designs are possible.

1.9   Invariant Parameters

Many of the parameters discussed in this chapter are invariant to lossless transformations. For example, cascading lossless networks to the input or output will not change the stability factor k or the power gains G_{ma}, G_{ms}, and U.

The unilateral gain U is also invariant to the common-lead. For example, a common-source, a common-gate, and a common-drain GaAs MESFET have the same unilateral gain. Since f_{max} is the upper frequency where unilateral gain is unity, this parameter is also invariant to the common-lead.

Other invariant parameters of interest are F_{min} (the minimum noise figure) and

$$N = R_n G_G \qquad (1.181)$$

Where the noise figure of a two-port can be written

$$F = F_{\min} + \frac{R_n}{G_G} |Y_G - Y_{\text{on}}|^2 \tag{1.182}$$

The four noise parameters of the two-port are:

$$F_{\min}$$
$$R_n = \frac{N}{G_G}$$
$$Y_{\text{on}} = G_{\text{on}} + jB_{\text{on}}$$

The $F_{\min}$ is invariant to lossless transformations and the common-lead. The N-parameter is only invariant to lossless transformations. Further discussion of low-noise design will be found in Chapter 3.

Bibliography

1.1 R. W. Anderson, "S-Parameter Techniques for Faster, More Accurate Network Design," *Hewlett Packard Application Note 95-1*, February 1967.

1.2 L. Besser, "Combine S-Parameters with Time-Sharing," *Electronic Design*, Vol. 16, August 1968, pp. 62–68.

1.3 G. E. Bodway, "Two Port Power Flow Analysis Using Generalized Scattering Parameters," *Microwave Journal*, Vol. 10, May 1967, pp. 61–69.

1.4 G. E. Bodway, "Circuit Design and Characterization of Transistors by Means of Three-Port Scattering Parameters," *Microwave Journal*, Vol. 11, May 1968, pp. 55–63.

1.5 R. S. Carson, *High Frequency Amplifiers*, Wiley, New York, 1975.

1.6 H. Fukui, "Available Power Gain, Noise Figure, and Noise Measure of Two-Ports and Their Graphical Representations," *IEEE Transactions on Circuit Theory*, Vol. CT-13, June 1966, pp. 137–142.

1.7 D. M. Kerns and R. W. Beatty, *Basic Theory of Waveguide Junctions and Introductory Microwave Network Analysis*, Pergamon, Elmsford, N.Y., 1967.

1.8 K. Kurokawa, "Power Waves and the Scattering Matrix," *IEEE Transactions on MTT*, Vol. MTT-13, March 1965, p. 194.

1.9 K. Kurokawa, *An Introduction to the Theory of Microwave Circuits*, Academic, New York, 1969.

1.10 R. Lane, "The Determination of Device Noise Parameters," *Proceedings of the IEEE*, Vol. 57, August 1969, pp. 1461–1462.

1.11 J. Lange, "Noise Characterization of Linear Two Ports in Terms of Invariant Parameters," *IEEE Journal on Solid State Circuits*, Vol. SC-2, June 1967, pp. 37–40.

1.12 J. G. Linvill and L. G. Schimpf, "The Design of Tetrode Transistor Amplifiers," *Bell Systems Technical Journal*, Vol. 35, July 1956, p. 818.

1.13 J. G. Linvill and J. F. Gibbons, *Transistors and Active Circuits*, McGraw-Hill, New York, 1961.

1.14 S. J. Mason, "Power Gain in Feedback Amplifiers," *IRE Transactions on Circuit Theory*, Vol. CT-1, June 1954, pp. 20–25.

1.15 J. M. Rollett, "Stability and Power Gain Invariants of Linear Two Ports," *IRE Transactions on Circuit Theory*, Vol. CT-9, March 1962, pp. 29–32.

1.16 "S-Parameter Design," *Hewlett Packard Application Note 154*, April 1972.

1.17 A. P. Stern, "Stability and Power Gain of Tuned Transistor Amplifiers," *Proceedings of the IRE*, Vol. 45, March 1957, pp. 335–343.

1.18 D. Woods, "Reappraisal of the Unconditional Stability Criteria for Active 2-Port Networks in Terms of S-Parameters," *IEEE Transactions on Circuits and Systems*, Vol. CAS-23, February 1976, pp. 73–81.

1.19 D. C. Youla, "On Scattering Matrices Normalized to Complex Port Numbers," *Proceedings of the IRE*, Vol. 49, No. 7, July 1961, p. 1221.

CHAPTER TWO

MICROWAVE TRANSISTOR S-PARAMETERS AND TUNING ELEMENTS

2.0 Introduction

The design of modern microwave amplifiers and oscillators requires impedance matching techniques. From a knowledge of the active two-port (or three-port) S-parameters, of microstrip transmission line elements, and of impedance matching techniques, the amplifier or oscillator design can be completed. These additional tools are presented in this chapter. In addition, an understanding of device physics is needed in order to properly dc bias the transistor to the required operating point.

The choice of transistors is usually between the silicon (Si) bipolar transistor and the GaAs MESFET. A comparison of the relative gain, noise figure, and power is summarized in Table 2.1. It should be noted that the GaAs MESFET has lower noise, higher gain, and higher power output. The primary disadvantage is higher $1/f$ noise, which can be significant for oscillator noise.

The higher output power of the GaAs MESFET is a direct result of the higher critical field and higher saturated drift velocity. The approximate power-frequency-squared limit is given by

$$Pf^2 \simeq \left(\frac{E_c v_s}{2\pi} \right)^2 \frac{1}{X_c} \qquad (2.1)$$

where E_c = effective electric field before avalanche breakdown
   v_s = drift velocity of carriers (electrons)
   X_c = device impedance level

and is a property of the material. Since the parameters E_c and v_s are higher for GaAs, the GaAs MESFET is intrinsically a higher-power device.

Table 2.1 1981 Comparison of Microwave Transistors

Parameters	GaAs MESFET				Si Bipolar Transistors		
	4 GHz	8 GHz	12 GHz	18 GHz	4 GHz	8 GHz	12 GHz
Gain (dB)	20	16	12	8	15	9	6
F_{min} (dB)	1.0	1.8	2.2	2.5	2.5	4.5	8
Power output (W)	25	8	4	1	6	2	0.25
Oscillator noise $1/f$ corner frequency	100 MHz				10 kHz		
Avalanche breakdown field E_{max} (V/cm)	4×10^5				3.5×10^5		
Saturated drift velocity v_s (cm/sec)	2×10^7				0.7×10^7		
1981 Pf^2 limit (W/sec^2)	10^{21}				10^{20}		
Theoretical Pf^2 limit (W/sec^2)	5×10^{21}				5×10^{20}		

If we include a correction factor for the geometry of the transistor, the effective values for GaAs are

$$X_c \simeq 1 \text{ ohm}$$

$$E_c \simeq \frac{E_{max}}{4} \simeq 10^5 \text{ V/cm}$$

$$v_s \simeq \frac{v_{sat}}{5} \simeq 4 \times 10^6 \text{ cm/sec}$$

$$Pf^2 \simeq 5 \times 10^{21} \text{ W/sec}^2$$

In 1980 the CW performance of 10 W at 10 GHz had already been achieved, giving

$$Pf^2 = 10 \times 10^{10} \times 10^{10} = 10^{21} \text{ W/sec}^2$$

The CW Si bipolar transistor will presently reach

$$Pf^2 \simeq 10^{20} \text{ W/sec}^2$$

with a theoretical limit of 5×10^{20} W/sec^2, an order of magnitude lower. The measured Pf^2 feature improves by about a factor of two under pulsed conditions. The advantages of Si are lower cost, higher thermal conductivity, and lower $1/f$ noise. The device limitations on frequency response are the transit time of electrical charge and the rate of change of electrical charge. These limitations will be discussed in this chapter.

2.1   The Si Bipolar Transistor

The Si bipolar transistor is shown in cross section in Fig. 2.1. The base current modulates the collector current of the transistor. Normally the CE or CB configuration is chosen for power gain. The emitter-base junction is forward biased, and the collector-base junction is reverse biased. The base current must move through a distributed RC line before reaching the active portion of the transistor, the emitter periphery. Although the distributed nature of the bipolar requires an effective value of r_b' and C_c, the figure of merit for the bipolar is

$$f_{max}^2 = \frac{f_t}{8 \pi r_b' C_c} \qquad (2.2)$$

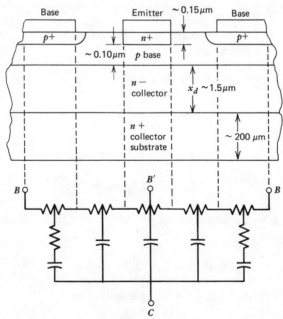

Figure 2.1   Bipolar transistor cross section (NPIN).

where f_{max} is the frequency at which unilateral gain becomes unity and f_t represents the delay time from emitter to collector (i.e., the transit time). The transit time is given by

$$\tau_{ec} = \tau_e + \tau_{eb} + \tau_{bc} + \tau_b + \tau_d + \tau_c \qquad (2.3)$$

where $\quad \tau_e =$ emitter delay due to excess holes in emitter

$\tau_{eb} =$ emitter-base capacitance charging time through emitter $=$

$$r_e' C_{Te} = \frac{kT}{qI_E} C_{Te}$$

$\tau_{bc} =$ base-collector capacitance charging time through emitter $=$

$$r_e' C_c$$

$\tau_b =$ base transit time

$\tau_d =$ collector depletion layer delay time $= \dfrac{X_d}{2v_s}$

$\tau_c =$ base-collector capacitance charging time through collector

Equation (2.3) usually reduces to

$$\tau_{ec} \simeq \tau_b + \tau_d \simeq 20 \text{ psec}$$

The frequency at which the common-emitter current gain reduces to unity is defined by f_t and is determined by the delay time from emitter to collector τ_{ec} according to the equation

$$f_t = \frac{1}{2\pi \tau_{ec}} \qquad (2.4)$$

For a high-quality Si bipolar transistor

$$f_t \simeq \frac{1}{2\pi(20)(10^{-12})} = 8 \text{ GHz}$$

$$r_b' C_c \simeq \frac{(15)(0.10) \times 10^{-12}}{3} \simeq 0.5 \text{ psec}$$

$$f_{max} \simeq \sqrt{\frac{8 \times 10^9}{8\pi(0.5) \times 10^{-12}}} \simeq 25 \text{ GHz}$$

The common-emitter and common-base equivalent circuits for a Si bipolar transistor chip are shown in Fig. 2.2. The bonding inductances to the base and emitter must also be included in the rf design, usually about 0.3 nH or less. Typical parameter values are given for a modern Si bipolar transistor (Hewlett Packard HXTR-2001) at $V_{CE} = 15$ V, $I_{CE} = 25$ mA.

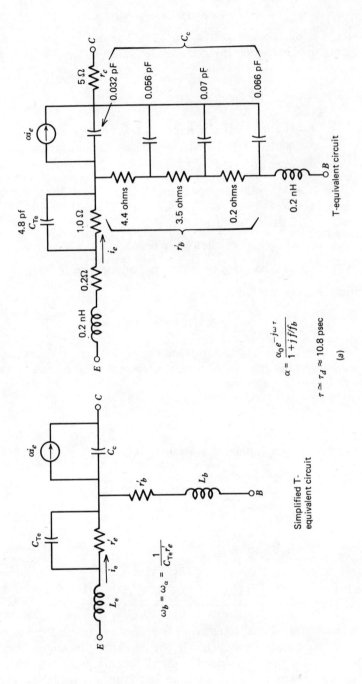

T-equivalent circuit

$$\alpha = \frac{\alpha_0 e^{-j\omega\tau}}{1 + jf/f_b}$$

$\tau \simeq \tau_d \approx 10.8$ psec

(a)

Simplified T-equivalent circuit

$$\omega_b = \omega_\alpha = \frac{1}{C_{Te} r'_e}$$

44

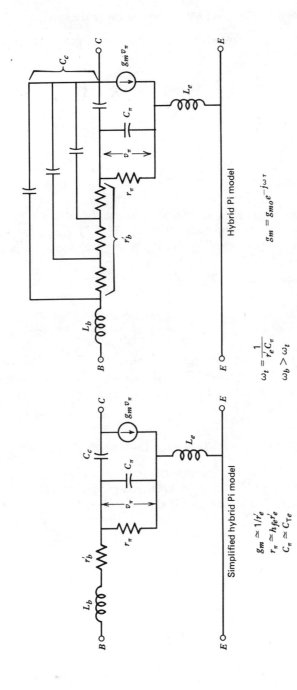

Figure 2.2 Equivalent circuits of bipolar transistor chip. (*a*) Common-base bipolar (HXTR-2001); (*b*) common-emitter bipolar.

Hybrid Pi model

$$g_m = g_{mo}e^{-j\omega\tau}$$

$$\omega_t = \frac{1}{r'_e C_\pi}$$

$$\omega_b > \omega_t$$

(*b*)

Simplified hybrid Pi model

$$g_m \simeq 1/r'_e$$

$$r_\pi \simeq h_{fe} r'_e$$

$$C_\pi \simeq C_{Te}$$

45

For the T-equivalent circuit in Fig. 2.2, the α is frequency-dependent and is given by

$$\alpha = \frac{\alpha_0}{1 + j\dfrac{f}{f_b}} \exp(-j2\pi f\tau) \qquad (2.5)$$

where $\alpha_0 \simeq 0.99$
$\qquad f_b = 22.7$ GHz = common-base cut off frequency
$\qquad \tau = 10.8$ psec

For optimum design of Si bipolar transistors, the parasitic resistances and capacitances must be minimized. In addition, (2.2)–(2.4) show that the minimum values of r_b', C_c, and τ_{ec} will give the maximum frequency of operation and therefore the maximum gain.

A fair approximation for the bipolar is

$$U \simeq G_{\mathrm{ma}} \simeq \left(\frac{f_{\max}}{f}\right)^2 \qquad (2.6)$$

which gives an estimate of the transistor gain. The G_{ma} is usually a few decibels lower than U in practice. The S-parameters of the transistor should be used to give a more accurate calculation of G_{ma}, using the equations in the previous chapter.

2.2 The Gallium Arsenide MESFET

The GaAs MESFET is more commonly used in microwave integrated circuit designs because of higher gain, higher output power, and lower noise figure in amplifiers. The higher gain is due to higher mobility of electrons (compared to silicon). The improvement in output power is due to the higher electric field and higher saturated drift velocity of the electrons [Eqn. (2.1)]. The lower noise figure is partially due to the higher mobility of the electron carriers. Moreover, fewer noise sources are present in the FET (no shot noise) as compared to the bipolar transistor. A disadvantage of the GaAs MESFET is the higher $1/f$ noise as compared to Si bipolar transistors.

Because of the superior properties of GaAs, an obvious question is, Where is the GaAs bipolar transistor? The difficulty lies in manufacturing an NPIN bipolar structure with GaAs. The base impurities diffuse faster than the emitter impurities, so a narrow diffused base with a high doping concentration has not been possible to manufacture with GaAs. In addition,

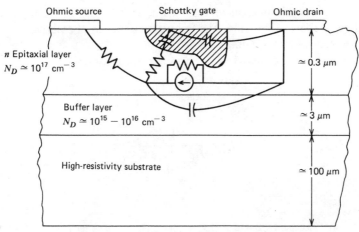

Figure 2.3 GaAs MESFET cross section.

the mobility of the p-type base region and the minority carrier lifetime in the base region are both relatively low. If these technological problems can be solved, the GaAs bipolar may become a useful microwave product.

The cross section of the GaAs MESFET (MEtal-Semiconductor Field-Effect Transistor) is shown in Fig. 2.3. The name MESFET has been adopted because of the similarity to MOSFET (Metal-Oxide Semiconductor Field-Effect Transistor). In Fig. 2.3 the electrons are drawn to the drain by a V_{DS} supply that accelerates the carriers to the maximum drift velocity, $v_s \simeq 2 \times 10^7$ cm/sec. The reverse bias of the Schottky-barrier gate allows the width of the channel to be modulated at a microwave frequency. Thus the majority carrier electrons are modulated by the input signal voltage applied across the input capacitance. There are several interesting contrasts between the FET and the bipolar, which are summarized in Table 2.2.

Table 2.2 Characteristics of Bipolar Transistor versus MESFET

Property	Common-Emitter Bipolar	Common-Source MESFET
Geometry	Vertical	Horizontal
Modulation	Base current	Gate voltage
Control signal	Current	Voltage
Frequency limitation	Base width	Gate length
Low-Frequency transconductance	High	Low

The frequency limitation of the FET is due to the gate length, which should be as short as possible. The frequency limits can be derived from the simplified hybrid π model in Fig. 2.4.

For the simplified model, the short-circuit current gain is

$$h_{21} = \frac{I_{\text{out}}}{I_{\text{in}}} = \frac{(g_m)v_c}{I_{\text{in}}} \tag{2.7}$$

$$I_{\text{in}} = \frac{V_{\text{in}}}{R_c + 1/j\omega C_{\text{gs}}} \tag{2.8}$$

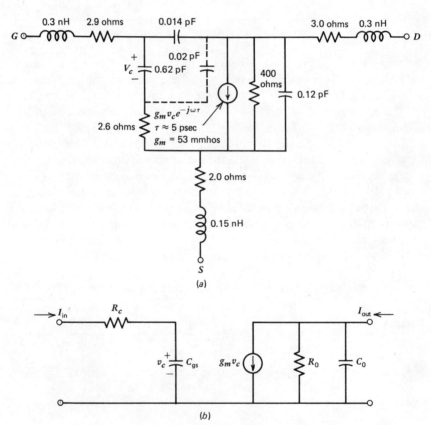

Figure 2.4 Equivalent circuit of GaAs MESFET (HFET-1001). (*a*) Hybrid pi model for $Z = 500 \ \mu\text{m}$, $L_g = 1.0 \ \mu\text{m}$ (HFET-1001) for $V_{\text{DS}} = 5.0$ V, $I_{\text{DS}} = I_{\text{DSS}}$; (*b*) simplified hybrid pi model.

At low frequencies:

$$I_{in} \simeq V_{in} j\omega C_{gs} \simeq v_c j\omega C_{gs} \qquad (2.9)$$

$$h_{21} \simeq \frac{g_m}{j\omega C_{gs}} \qquad (2.10)$$

$$|h_{21}| = \frac{f_t}{f} = \frac{g_m}{2\pi C_{gs}} \frac{1}{f} \qquad (2.11)$$

Thus the frequency where the short circuit current gain becomes unity is

$$f_t = \frac{g_m}{2\pi C_{gs}} \qquad (2.12)$$

which is an important figure of merit for the GaAs MESFET. The unilateral gain of the FET may be simply calculated from the y-parameters in Fig. 2.4.

$$y_{11} = \frac{1}{R_c + \dfrac{1}{j\omega C_{gs}}} \qquad (2.13)$$

$$y_{21} = g_m \qquad (2.14)$$

$$y_{22} = \frac{1}{R_o} + j\omega C_o \qquad (2.15)$$

$$y_{12} = 0 \qquad (2.16)$$

From (1.162)

$$U = \frac{|y_{21}|^2}{4 \operatorname{Re} y_{11} \operatorname{Re} y_{22}} \qquad (2.17)$$

$$U = \frac{1}{4} \frac{1}{f^2} \left(\frac{g_m}{2\pi C_{gs}} \right)^2 \frac{R_o}{R_c}$$

$$= \left(\frac{f_{max}}{f} \right)^2 \qquad (2.18)$$

Thus f_{max} is given by

$$f_{max} = \frac{f_t}{2} \sqrt{\frac{R_o}{R_c}} \qquad (2.19)$$

A high-gain FET requires a high f_t, a high output-resistance, a low input-resistance, and minimum parasitic elements. Under normal bias conditions the device is biased for maximum drift velocity of the electron carriers (about 3kV/cm), so we have

$$f_t = \frac{g_m}{2\pi C_{gs}} = \frac{1}{2\pi\tau} = \frac{v_s}{2\pi L_g} \qquad (2.20)$$

This equation shows the importance of short gate length L_g. Another interesting figure of merit for the FET is the g_m per unit gate periphery (Z), given by

$$C_{gs} = \frac{\varepsilon A}{d} = \varepsilon \frac{L_g Z}{d} \qquad (2.21)$$

$$g_m = \frac{v_s C_{gs}}{L_g} = \frac{v_s \varepsilon Z}{d} \qquad (2.22)$$

$$\frac{g_m}{Z} = \frac{v_s \varepsilon}{d} \qquad (2.23)$$

For a typical ($Z = 500\,\mu\text{m}$ gate) FET, this parameter is

$$\frac{g_m}{Z} = \frac{2\times 10^7 \text{ cm/sec } 10^{-12} \text{ Fd/cm}}{0.3\,\mu\text{m}}$$

$$= 67\,\mu\text{mhos}/\mu\text{m} = 0.067 \text{ mmhos}/\mu\text{m}$$

$$g_m \simeq 0.067 \times 500 = 33 \text{ mmhos}$$

which is in good agreement with measurements. Scaling the device larger in Z increases the transconductance, but the f_t and gain remain constant with scaling if parasitics are negligible.

The velocity saturating effect of the GaAs MESFET has several interesting consequences. Referring to Fig. 2.5, we see that the channel can be considered to be two regions: a low-field region with a constant number of carriers and a high-field region with a "constant" velocity, which is discussed later. Since the current continuity is required

$$\frac{I_{DS}}{A} = qn(x)v(x) \qquad (2.24)$$

where

$$n(x)_{\max} = N_D$$

$$v(x)_{\max} = v_{\text{sat}} = v_s$$

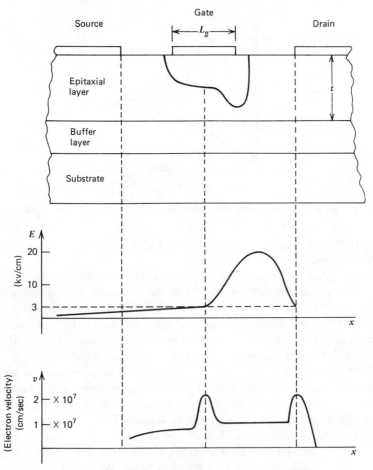

Figure 2.5   GaAs MESFET at high electric field.

the number of carriers must increase above N_D in Region II. This causes an
electron accumulation at the drain edge of the channel, followed by an
electron depletion. In effect, a charge dipole occurs at the drain edge of the
channel, which is a very small capacitive effect in the model ($\simeq 0.02$ pF in
Fig. 2.4).

In GaAs, the electron carriers will slow down at an electric field greater
than 3 kV/cm. The electrons move from a high-mobility state to a low-
mobility state in about 1 psec, and thus the velocity of the carriers reaches a
peak and slows down in the middle of the channel. The velocity versus
electric field of GaAs and Si is shown in Fig. 2.6. The change in velocity of

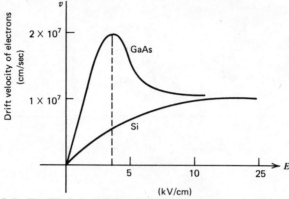

Figure 2.6   Equilibrium electron drift velocity versus electric field.

the carriers in GaAs is the cause of the Gunn effect in Gunn diodes or TEOs, (Transfer Electron Oscillators).

In short-channel devices (less than 3 μm), a nonequilibrium velocity field characteristic must be considered. When the electrons enter the high-field region, they are accelerated to a higher velocity. This effect can cause peak velocities of about 4×10^7 cm/sec, which relax to 1×10^7 cm/sec after traveling about 0.5 μm. The overshoot in velocity reduces the transit time and shifts the dipole charge to the right of the channel.

The high-frequency gain of the GaAs MESFET is maximized by achieving the minimum gate length without introducing excessive device parasitics. Computer studies have shown that the R_G series gate resistance increases and R_o decreases as the gate length is shortened. The practical limit is $L_g/t > 1$, which implies a thin channel and therefore higher channel doping. As a result of breakdown considerations, the maximum channel doping is 4×10^{17} cm^{-3}, and about 2×10^{17} cm^{-3} in practical devices. Thus modern 0.5 μm gate GaAs MESFETs are probably within a factor of 3 of the highest $f_{\max}$ that can be achieved from the present device structure.

Evaluating $f_{\max}$ for the device shown in Fig. 2.4 gives

$$f_t = \frac{0.053}{2\pi(0.62)} = 13.6 \text{ GHz}$$

$$f_{\max} = \frac{13.6}{2}\sqrt{\frac{400}{7.5}} \simeq 50 \text{ GHz}$$

These frequencies are typical of modern 1.0 μm GaAs MESFETS.

The dual-gate GaAs MESFET is simply two adjacent gates with a cascade connection normally used (CS followed by CG). The cross section

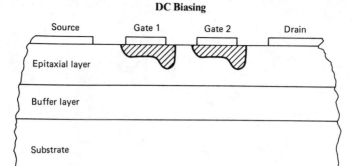

Figure 2.7 Dual-gate GaAs MESFET cross section.

in Fig. 2.7 is typical of the dual-gate transistor. The second gate can be used for AGC by varying the dc voltage at gate 2. As an amplifier, the dual gate FET has higher gain with gate 2 rf grounded. Many other applications are possible for the dual-gate transistor, including oscillators, mixers, and multipliers.

2.3 DC Biasing

Before considering the rf design of an amplifier or oscillator, the dc biasing point should be established. Poor performance in a final design can often be traced to improper dc biasing.

The common-emitter or common-source biasing configuration is discussed here. Notice that the dc biasing is completely independent of the rf two-port configuration. For example, a transistor may operate as an amplifier in the common-gate mode, but the dc biasing circuit is in the common-source mode. The dc collector and drain characteristics of the bipolar and FET are given in Fig. 2.8.

The biasing point depends upon the application, but some typical guidelines are listed in Table 2.3. The high-power biasing point is determined by the limitations of the particular transistor and its safe operating region. The definitions of Class A, AB, B, and C are given graphically in Fig. 2.9.

The safe operating region for the bipolar is determined by

1 Maximum collector current.

2 Maximum collector-emitter voltage.

3 Secondary breakdown.

4 Maximum power dissipation (maximum junction temperature $\simeq 200°C$ for Si).

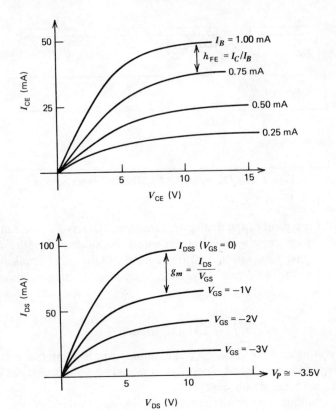

Figure 2.8 dc characteristics of Si bipolar transistors and GaAs MESFETs.

Table 2.3   Transistor Bias Points

Application	Si Bipolar	GaAs MESFET ($I_{DSS} \simeq 80$mA)
Low noise	$V_{CE} = 10$ V, $I_{CE} = 3$ mA	$V_{DS} = 3.5$ V, $I_{DS} = 10$ mA
High gain	$V_{CE} = 10$ V, $I_{CE} = 10$ mA	$V_{DS} = 5$ V, $I_{DS} = 80$ mA
High output power	$V_{CE} \geq 20$ V, $I_{CE} = 25$ mA	$V_{DS} \geq 10$ V, $I_{DS} = 40$ mA
Low distortion	$V_{CE} \geq 20$ V, $I_{CE} = 25$ mA	$V_{DS} \geq 10$ V, $I_{DS} = 40$ mA
Class B	$V_{CE} \geq 20$ V, $I_{CE} = 0$	$V_{DS} \geq 8$ V, $I_{DS} = 0$
Class C	$V_{CE} \geq 28$ V, $I_{CE} = 0$	Not used

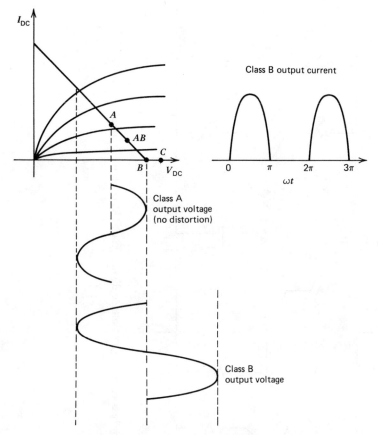

Figure 2.9   High-power operation of transistors.

For the GaAs MESFET, the safe operating region is determined by

1   Maximum drain current.
2   Maximum drain-source voltage.
3   Maximum power dissipation (maximum junction temperature $\simeq 175°C$ for GaAs presently).
4   Maximum input power to gate.

For GaAs the junction temperature is limited by a possible chemical reaction to 175°C for high reliability. Moreover, lower thermal conductivity occurs for this material, so excellent heat sinking must be provided.

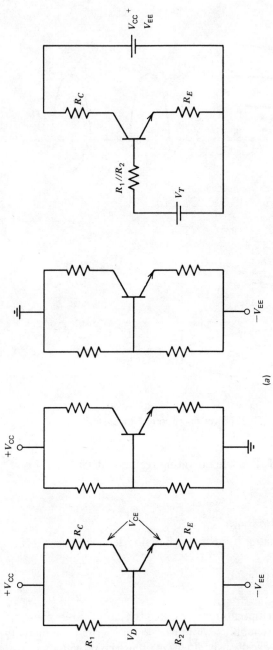

Figure 2.10 Bipolar transistor bias circuits. (*a*) Voltage divider bias; (*b*) collector feedback bias; (*c*) active bias.

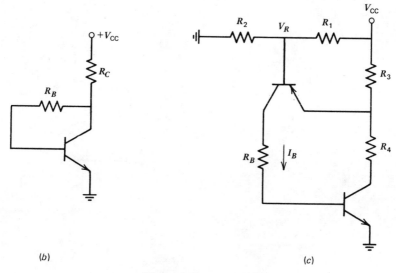

(b)   (c)

Figure 2.10 (*Continued*)

Some biasing circuits are tabulated in Fig. 2.10 for bipolars and in Fig. 2.11 for MESFETS. For Si bipolars, the emitter-base is forward biased at about 0.7 V and the collector-base is reverse biased at, typically, 10 V. For the FET, the gate to source is reverse biased and the drain to source is biased to the "saturation region," greater than 3 V, where the electrons travel at saturated drift velocity.

For the voltage-divider circuits, the design equations are

$$V_{CC} + V_{EE} = I_C R_C + V_{CE} + I_E R_E \qquad (2.25)$$

$$V_D = 0.7 + I_E R_E \qquad (2.26)$$

By making

$$R_E \gg \frac{R_B}{h_{FE}} = \frac{R_1 R_2}{R_1 + R_2} \cdot \frac{1}{h_{FE}}$$

the bias circuit becomes independent of h_{FE}.

For collector feedback bias,

$$V_{CC} = I_C R_C + V_{CE} \qquad (2.27)$$

$$V_{CE} = I_B R_B + 0.7 \qquad (2.28)$$

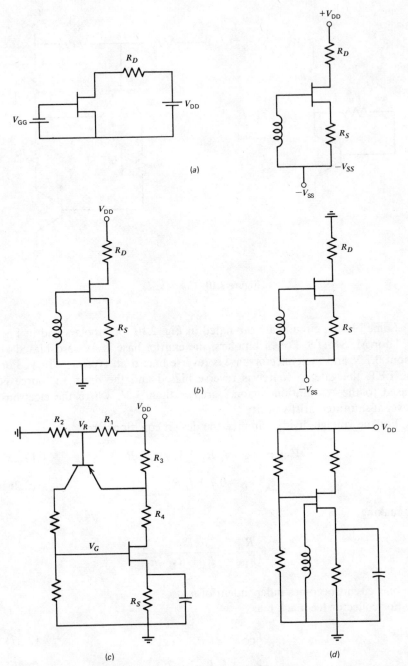

Figure 2.11 MESFET bias circuits. (*a*) Two-supply bias; (*b*) single-supply bias; (*c*) active bias; (*d*) dual-gate bias.

The feedback effect in this circuit can be understood by considering an increase in h_{FE}. This will increase I_C, which reduces V_{CE}, which reduces I_B, which stabilizes the collector current to a lower value than expected. Thus the collector feedback circuit has a self-stabilizing effect.

For active bias, a PNP transistor is used to maintain the operating point of the microwave NPN bipolar transistor. A voltage reference is set by

$$V_R = \frac{R_2}{R_1 + R_2} V_{\text{CC}} \tag{2.29}$$

The collector current is set by R_3 according to

$$V_{\text{CC}} = I_C R_3 + 0.7 + V_R \tag{2.30}$$

The collector-emitter voltage is set by

$$V_{\text{CC}} = I_C (R_3 + R_4) + V_{\text{CE}} \tag{2.31}$$

Notice that R_4 is not required. Thus three resistors, R_1, R_2, and R_3, and a bipolar will stabilize the operating point by supplying the base current to maintain a constant collector current to the bipolar transistor. This is a stable bias circuit with temperature, since V_{BE} of the PNP bipolar changes only 2.5 mV/°C.

The bias circuits for FETS are given in Fig. 2.11. Normally the single-supply circuits are preferred with the reverse bias gate-source given by

$$V_{\text{GS}} = I_{\text{DS}} R_S \tag{2.32}$$

The first two-supply circuit may be required when dc efficiency is important. The active bias circuit is similar to the bipolar case, only now the PNP bipolar regulates the gate voltage to maintain a constant operating point. The single-supply dual-gate circuit is shown to indicate the simplicity of dual-gate bias.

2.4 Microstrip Transmission Lines

Modern microwave integrated circuits with active transistors are designed using lossless matching or tuning elements. These elements may be any of the following in practical circuit designs:

1 Lumped inductors.
2 Lumped capacitors.
3 Series microstriplines.

4 Shunt open-circuited microstripline stubs.

5 Shunt short-circuited microstripline stubs.

The designs are usually a hybrid combination of these lossless tuning elements. Before we consider the impedance matching techniques in the next section, the important electrical parameters of the microstripline must be considered.

The basic microstripline geometry and field pattern are given in Fig. 2.12. This transmission line may be visualized as a distorted coaxial line, with the top strip as the center conductor and the ground plane transformed to a flat plane. The electrical behavior is determined by the distributed inductance and capacitance of the line (l and c). The important physical and electrical parameters are defined in Table 2.4 for microstripline. The important physical parameters are line width w, substrate thickness h, and substrate dielectric constant ε_r. Refer to Section 1.1 for our earlier discussions of transmission lines.

The transmission mode is assumed to be quasi-TEM, that is, there is no electric field or magnetic field in the direction of propagation. The electrical properties of the microstripline can be calculated from finding the effective dielectric constant of the transmission line. A portion of the electric field lines is in the dielectric, and a few fringe electric field lines occur in the air. The distribution of the electric field lines is described by the filling factor q, which approaches unity when all of the electric field lines are in the substrate. The effective or microstrip dielectric constant ε_r' is related to the filling factor q by

$$\varepsilon_r' = q\varepsilon_r + (1 - q)\varepsilon_0 \tag{2.33}$$

$$q = \frac{\varepsilon_r' - \varepsilon_0}{\varepsilon_r - \varepsilon_0} \leq 1 \tag{2.34}$$

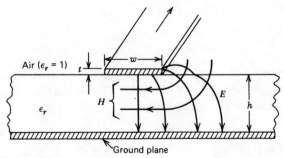

Figure 2.12 Microstripline geometry (w = width; h = height).

Table 2.4   Microstripline Parameters

Physical		Electrical	
w	line width	Z_0	characteristic impedance
h	substrate thickness	α	attenuation constant
ε_r	substrate dielectric constant		(nepers/length)
		β	phase constant (radians/
t	metal thickness		length)
σ_m	metal conductivity	v_{ph}	phase velocity
σ_d	dielectric conductivity	ε_r'	effective dielectric
			constant
		λ_g	guide wavelength
		q	filling factor
		r	resistance per unit length
		l	inductance per unit length
		g	conductivity per unit length
		c	capacitance per unit length

If we define a dimensionless dielectric constant by

$$k = \frac{\varepsilon_r}{\varepsilon_0} \tag{2.35}$$

$$k' = \frac{\varepsilon_r'}{\varepsilon_0} \tag{2.36}$$

these equations can be written

$$k' = qk + (1 - q) \tag{2.37}$$

$$q = \frac{k' - 1}{k - 1} \leq 1 \tag{2.38}$$

For wide lines, q approaches unity, ε_r' approaches ε_r, and k' approaches k. For narrow lines, q approaches $\frac{1}{2}$, ε_r' is the average of ε_r and ε_0, and k' is the average of k and unity.

The effective dielectric constant reduces the phase velocity and reduces the characteristic impedance according to the following relations:

$$v_{\mathrm{ph}} = \frac{1}{\sqrt{lc}} = \frac{c_0}{\sqrt{k'}} \tag{2.39}$$

$$Z_0 = \sqrt{\frac{l}{c}} = \frac{Z_{0a}}{\sqrt{k'}} \tag{2.40}$$

where c_0 is the velocity of light and Z_{0a} is the characteristic impedance for an air dielectric. The velocity of the wave is reduced by the square root of the effective dielectric constant, but not as slowly as if all of the wave is in the dielectric. The characteristic impedance is reduced by the increased capacitance of the line compared to the air line. The guide wavelength is also reduced according to the relations

$$\lambda_0 = \frac{c_0}{f} \tag{2.41}$$

$$\lambda_g = \frac{v_{\mathrm{ph}}}{f} = \frac{c_0}{f\sqrt{k'}} \tag{2.42}$$

Since c_0 (the velocity of light) is 3×10^{10} cm/sec, a useful relation at microwave frequencies is

$$\lambda_g = \frac{30}{f_{\mathrm{GHz}}\sqrt{k'}} \text{ cm} \tag{2.43}$$

Notice that the phase velocity, characteristic impedance, and guide wavelength are all dependent on the effective relative dielectric constant k'.

A brief review of transmission line theory will aid in understanding the propagation constant of the line. The distributed constants of the line are

r	Resistance per unit length
l	Inductance per unit length
g	Conductance per unit length
c	Capacitance per unit length

The distributed model of the line is given in Fig. 2.13.

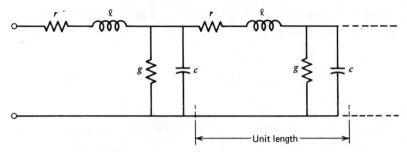

Figure 2.13 Distributed transmission line model.

The propagation constant of the voltage and current waves is $e^{\pm \gamma x}$ [see (1.1)]

$$\gamma = \alpha + j\beta = \sqrt{(r+j\omega l)(g+j\omega c)} \qquad (2.44)$$

$$Z_0 = \sqrt{\frac{r+j\omega l}{g+j\omega c}} = \frac{V_{inc}}{I_{inc}} = \frac{V_{ref}}{I_{ref}} \qquad (2.45)$$

For lossless lines these reduce to

$$\gamma \simeq j\beta = j\omega\sqrt{lc} = j\frac{\omega}{v_{ph}} \qquad (2.46)$$

$$Z_0 \simeq \sqrt{\frac{l}{c}} \qquad (2.47)$$

The concept of attenuation can be explained from (1.2) and (1.3), and Fig. 1.3. Consider a line terminated in characteristic impedance Z_0, so that the reflected waves are zero ($V_{ref} = I_{ref} = 0$). For this line at any point x, the load voltage and load current become

$$V_L = V(x) = V_{inc}e^{-\alpha x}e^{-j\beta x} \qquad (2.48)$$

$$I_L = I(x) = I_{inc}e^{-\alpha x}e^{-j\beta x} \qquad (2.49)$$

The ratio of load voltage to incident voltage and the ratio of load current to incident current are given by

$$\left|\frac{V_L}{V_{inc}}\right| = e^{-\alpha x} \qquad (2.50)$$

$$\left|\frac{I_L}{I_{inc}}\right| = e^{-\alpha x} \qquad (2.51)$$

For a real characteristic impedance, the ratio of power delivered to the load to incident power becomes

$$\frac{P_L}{P_{\text{inc}}} = \left| \frac{V_L}{V_{\text{inc}}} \right| \left| \frac{I_L}{I_{\text{inc}}} \right| = e^{-2\alpha x} \tag{2.52}$$

Thus the attenuation constant α reduces the power delivered to the load. This can also be written

$$\alpha x = \frac{1}{2} \ln \frac{P_{\text{inc}}}{P_L} \tag{2.53}$$

where the units of this equation are nepers (dimensionless), so that α is in nepers per unit length.

Since attenuation (A) is often given in decibels,

$$A = 10 \log \frac{P_{\text{inc}}}{P_L} \tag{2.54}$$

$$= 10 \log e^{2\alpha x}$$

$$= 20 \alpha x \log e$$

$$= 8.686 \alpha x \tag{2.55}$$

Thus 1 dB is 8.686 Np, with both quantities dimensionless ratios of power.

The effect of losses can be found at high frequencies by expanding (2.44) as follows (assuming $rg \ll \omega^2 lc$ for low-loss lines):

$$\alpha + j\beta = \sqrt{(rg - \omega^2 lc) + j\omega(lg + rc)}$$

$$\simeq j\omega\sqrt{lc} \left(1 + \frac{j\omega(lg + rc)}{-\omega^2 lc} \right)^{1/2} \tag{2.56}$$

Using the binomial expansion theorem

$$(1 + x)^{1/2} = 1 + \frac{1}{2}x + \frac{1/2(-1/2)}{2!}x^2 + \cdots \tag{2.57}$$

gives

$$\alpha + j\beta \simeq j\omega\sqrt{lc}\left[1 + \frac{1}{2}\frac{j\omega(lg + rc)}{-\omega^2 lc} - \cdots\right]$$

$$= j\omega\sqrt{lc} + \frac{lg + rc}{2\sqrt{lc}} \tag{2.58}$$

Thus, the attenuation constant consists of two terms that can be associated with conductor losses and dielectric losses (in units of nepers per length):

$$\alpha = \alpha_c + \alpha_d = \frac{r}{2}\sqrt{\frac{c}{l}} + \frac{g}{2}\sqrt{\frac{l}{c}}$$

$$= \frac{r}{2Z_0} + \frac{gZ_0}{2} \tag{2.59}$$

The conductor losses are usually dominant for microstripline. For wide lines, the following approximation for conductor losses is valid (refer to Fig. 2.14). At high frequencies the current flows only in the skin depth, so the resistance per unit length of the top line is

$$r_1 = \frac{\rho}{A} = \frac{\rho}{\delta w} \tag{2.60}$$

The sheet resistance in ohms per square is

$$R_S = \sqrt{\pi f \mu \rho} = \frac{\rho}{\delta} \tag{2.61}$$

$$r_1 = \frac{R_S}{w} \tag{2.62}$$

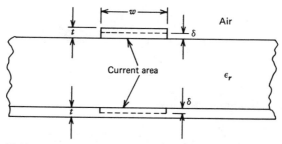

Figure 2.14 Uniform current approximation for microstrip line ($w =$ width).

If we assume equal contributions from top conductor and ground plane, the total conductor attenuation is

$$\alpha_c = \frac{r}{2Z_0} = \frac{R_S}{Z_0 w} \tag{2.63}$$

Using 1 dB = 8.686 Np, this is often written

$$\alpha_c = \frac{8.686 R_S}{Z_0 w} \tag{2.64}$$

where the units are decibels per unit length.

For dielectric losses, the loss tangent is usually given as a measure of the conductance losses by

$$\tan \delta_d = \frac{g}{\omega c} = \frac{\sigma_d}{\omega \varepsilon_r} \tag{2.65}$$

From (2.59)

$$\alpha_d = \frac{g Z_0}{2} = \frac{\omega c Z_0}{2} \tan \delta_d$$

$$= \frac{\omega \sqrt{lc}}{2} \tan \delta_d \tag{2.66}$$

Using

$$\sqrt{lc} = \sqrt{\mu \varepsilon_r} \tag{2.67}$$

we see that the dielectric loss of a uniformly loaded line ($q = 1$) is

$$\alpha_{du} = \frac{\omega}{2} \sqrt{\mu \varepsilon_r} \frac{\sigma_d}{\omega \varepsilon_r} = \sqrt{\frac{\mu}{\varepsilon_r}} \cdot \frac{\sigma_d}{2} \tag{2.68}$$

The filling factor gives

$$\sigma_d' = q \sigma_d \tag{2.69}$$

$$\varepsilon_r' = q \varepsilon_r + (1 - q) \varepsilon_0 \tag{2.70}$$

The microstripline dielectric loss becomes

$$\alpha_d = \alpha_{du} q \sqrt{\frac{k}{k'}} = \sqrt{\frac{\mu_0}{\varepsilon_0}} \frac{\sigma_d}{2} \frac{q}{\sqrt{k'}} \tag{2.71}$$

In units of decibels per unit length this becomes

$$\alpha_d = \frac{4.34q}{\sqrt{k'}} \sqrt{\frac{\mu_0}{\varepsilon_0}} \, \sigma_d \qquad (2.72)$$

From (2.64) and (2.72), the attenuation in units of decibels per unit length is

$$\alpha = \alpha_c + \alpha_d$$

$$= \frac{8.686\sqrt{\pi\mu\rho}}{Z_0 w} \sqrt{f} + \frac{4.34q}{\sqrt{k'}} \sqrt{\frac{\mu_0}{\varepsilon_0}} \, \sigma_d \qquad (2.73)$$

Using (2.65) for σ_d in the second term gives

$$\alpha = \frac{8.68\sqrt{\pi\mu\rho}}{Z_0 w} \sqrt{f} + 27.3q\sqrt{\frac{k}{k'}} \, \frac{\tan\delta_d}{v_{\text{ph}}} f \qquad (2.74)$$

In units of decibels per unit length, the conductor attenuation factor increases with the square root of frequency and the dielectric attenuation factor increases in proportion to frequency (for $\tan\delta_d$ constant with frequency). For low-loss dielectrics, the conductor attenuation factor is usually dominant. Using (2.42) to convert to guide wavelength, we find that the attenuation factor becomes

$$\alpha_c = \frac{8.686\sqrt{\pi\mu\rho}}{Z_0 w \sqrt{f}} \, \frac{c_0}{\sqrt{k'}\lambda_g} \, \frac{\text{dB}}{\text{length}} \qquad (2.75)$$

For a fixed length of line ℓ, the attenuation in units of decibels per guide wavelength becomes

$$\alpha_c \ell = \frac{8.686\sqrt{\pi\mu\rho} \, c_0}{Z_0 w \sqrt{k'} \sqrt{f}} \, \frac{\text{dB}}{\lambda_g} \qquad (2.76)$$

which indicates that the attenuation is decreasing with frequency. This can be understood by realizing that the guide wavelength is also decreasing with frequency.

The microstripline attenuation is dominated by conductor losses. The attenuation in decibels per centimeter is increasing in proportion to $\sqrt{f}$, while the attenuation in decibels per guide wavelength is decreasing with $\sqrt{f}$. Another important empirical factor in microstripline attenuation is surface roughness, which increases the surface resistance of the metal conductors. This effect is used to explain the measurement of higher attenuation than predicted by (2.75).

The important substrate materials for microstripline are summarized in Table 2.5. The most common choice is alumina because of low cost, but surface roughness is also an important practical consideration.

The solutions for the microstripline characteristic impedance and effective dielectric constant are given in Figs. 2.15 and 2.16. These are the basic design curves for microstripline low-loss passive circuits.

The Q of microstripline is calculated from

$$Q = \frac{\beta}{2\alpha} \tag{2.77}$$

Using 1 dB = 8.686 Np gives

$$Q = \frac{2\pi}{\lambda_g} \frac{1}{2\alpha} = \frac{8.686\pi}{\alpha\lambda_g} \, dB$$

$$= \frac{27.3}{\alpha} \frac{dB}{\lambda_g} \tag{2.78}$$

Thus the Q of microstripline resonators will increase with frequency, since α

Table 2.5   Microstripline Dielectrics

Material	$k = \dfrac{\varepsilon_r}{\varepsilon_0}$	Comment
Alumina Al_2O_3	9.6–10	Low cost
Sapphire single crystal Al_2O_3	9.9 11.6	Expensive and Anisotropic
Fused silica (quartz)	3.78	Fragile
Beryllium oxide BeO	6.0	High thermal conductivity
Duroid	2.56	
Teflon-Fiberglass	2.32	
GaAs ($\rho = 10^7$ ohms cm)	12.3	Fragile
Silicon ($\rho = 10^3$ ohms cm)	11.7	Low resistivity

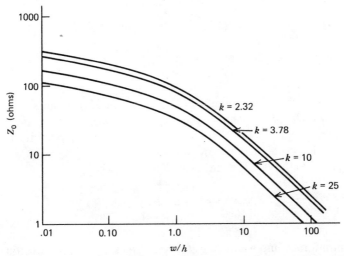

Figure 2.15 Microstripline characteristic impedance versus width/height (w/h) and dielectric constant.

in decibels per guide wavelength is decreasing. The dielectric loss and radiation losses should also be included for completeness. The dielectric loss is normally negligible except for semiconductor substrates such as silicon in which the conductivity of the substrate is significant.

The Q of an open-circuited resonator is thus modified to include conductor loss, dielectric loss, and radiation loss by

$$\frac{1}{Q_t} = \frac{1}{Q_c} + \frac{1}{Q_d} + \frac{1}{Q_r} \tag{2.79}$$

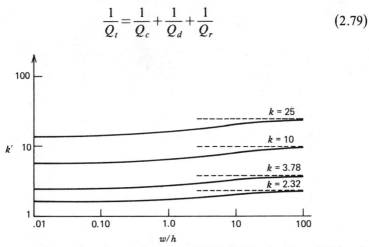

Figure 2.16 Effective dielectric constant versus width/height (w/h) and dielectric constant.

where

$$Q_c = \frac{27.3}{\alpha_c} \tag{2.80}$$

$$Q_d = \frac{27.3}{\alpha_d} \tag{2.81}$$

Radiation loss becomes significant for low-impedance lines, thick substrates, low dielectric constants, and high frequencies.

Another important effect for microstripline is the possibility of a TM surface mode for high dielectric substrates. The frequency at which this mode couples tightly to the microstripline mode is

$$f_c = \frac{c_0}{4h\sqrt{k-1}} \tag{2.82}$$

This limitation is important only for very thick or high k substrates. For 25-mil alumina, $f_c \simeq 50$ GHz.

The presence of the TM surface mode will cause dispersion of the line or variation of the effective dielectric constant with frequency. This causes the effective dielectric constant to increase with frequency, approaching the substrate dielectric constant. The effect can be explained by the coupling of the quasi-TEM mode to the lowest-order TM mode at frequencies where the phase velocities of these modes become comparable. A simple empirical relation for this effect is

$$v_{ph} = \frac{c_0 \left(\sqrt{k'} f_n^2 + \sqrt{k} \right)}{\sqrt{kk'} \left(f_n^2 + 1 \right)} = \frac{c_0}{\sqrt{k'(f)}} \tag{2.83}$$

$$f_n = \frac{f}{f_c} = \frac{4h\sqrt{k-1}}{\lambda_0} \tag{2.84}$$

In summary, the important electrical parameters for microstripline design are the characteristic impedance Z_0, the guide wavelength λ_g, and the attenuation constant α. The effective dielectric constant of the line (ε_r' or k') is geometry-dependent and frequency-dependent. Usually the attenuation is dominated by conductor losses and increasing with $\sqrt{f}$ in decibels per unit length, but decreasing with $\sqrt{f}$ in decibels per guide wavelength.

For the first-order design, a quasi-TEM solution is used, which gives the results in Figs. 2.15 and 2.16, the design curves for microstripline circuits. These curves ignore dispersion and radiation effects; nevertheless, they are very useful for normal 25-mil alumina microstripline circuits with $Z_0 = 30$ to

110 ohms below 18 GHz. Junction effects caused by cascading microstrip-line elements are also ignored in the first-order design.

2.5 Impedance Matching

An important design tool in amplifier and oscillator design is the concept of impedance matching. All values of impedance can be plotted in the reflection coefficient plane, where $|\Gamma| < 1$ for positive resistance loads. Recalling (1.9) and Fig. 1.3, we see that the impedance at any point on a transmission line is

$$\frac{Z(z)}{Z_0} = \frac{1 + \Gamma_L e^{-2\gamma z}}{1 - \Gamma_L e^{-2\gamma z}} = \frac{1 + \Gamma(z)}{1 - \Gamma(z)} \qquad (2.85)$$

$$\Gamma(z) = \Gamma_L e^{-2\gamma z} = \frac{Z(z) - Z_0}{Z(z) + Z_0} \qquad (2.86)$$

For any value of the reflection coefficient, a corresponding load impedance $Z(z)$ is found for the transmission line of characteristic impedance Z_0. For $|\Gamma_L| < 1$, only positive resistance values can occur. The polar plot of $|\Gamma_L| < 1$ is the Smith Chart plane. This chart can be visualized as both a plot of $\Gamma_L e^{2\gamma z}$ and a plot of $Z(z)$, the load impedance viewed at reference plane z. Notice that the phase shift as one moves toward the generator is $e^{-2\gamma z}$ ($\simeq e^{-j2\beta x}$ for lossless lines). Thus clockwise rotation on the Smith Chart plane is movement toward the generator.

Several important points on the Smith Chart are given in Fig. 2.17. Both the impedance plane and the admittance plane are shown. For microstrip-line shunt elements, the admittance plane should be used, since these are shunt matching elements. Examples of lossless shunt matching elements are given in Table 2.6 for understanding the technique of impedance (admittance) matching. These elements should be considered lossless matching elements.

Impedance matching normally involves using lossless matching elements to move from a point on the impedance (or admittance) plane to the center of the chart where $Z/Z_0 = Y/Y_0 = 1$. This movement should be done with as few lossless elements as possible, usually two, which should have the minimum length if microstriplines. The matching circuit could be a combination of lumped and distributed microstripline elements. Since an infinite number of solutions are possible, it is important to find several simple solutions so that practical considerations can be used to select the best design.

$$\Gamma = 1 \angle 0° \qquad \text{open circuit}$$

$$\Gamma = 1 \angle 180° \qquad \text{short circuit}$$

$$\Gamma = 1 \angle 90° \qquad \text{inductor}$$

$$\frac{Z}{Z_0} = \frac{1 + j1}{1 - j1} = j1$$

$$\Gamma = 1 \angle -90° \qquad \text{capacitor}$$

$$\frac{Z}{Z_0} = \frac{1 - j1}{1 + j1} = -j1$$

$$\Gamma = 0 \quad Z = Z_0$$

(a) Important Points on the Smith Chart

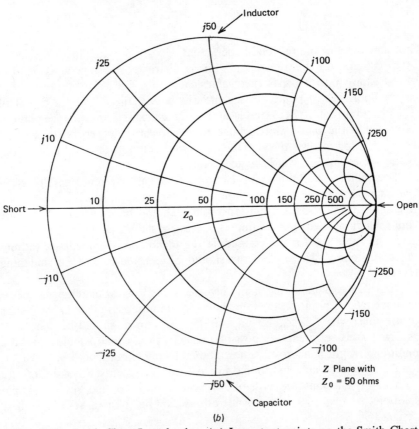

(b)

Figure 2.17 Smith Chart Introduction. (a) Important points on the Smith Chart. (b) Coordinates in ohms. Z plane with $Z_0 = 50$ ohms. (c) Y plane normalized to Y_0.

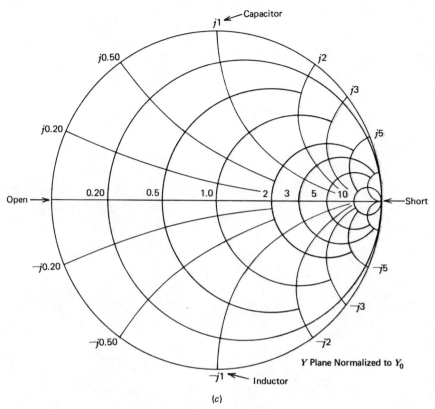

(c)

Figure 2.17 (*Continued*)

The Smith Chart is a useful tool for impedance matching. This is a circular chart that plots all values of Z_L or Y_L for positive values of R or G. Figure 2.18 shows the Smith Chart and a value of $Z_L/Z_0 = 0.5 + j0.5$. Also plotted are Z_L^*/Z_0, Y_L/Y_0, and Y_L^*/Y_0 which gives a rectangle on the Smith Chart. The usual impedance matching problem is to match Z_L to the center of the chart ($Z_0 = Z_G$), which gives a zero reflection coefficient and maximum power transfer from the generator to the load. If Y_L is matched to the

Table 2.6 Shunt Microstripline Matching Elements

Length $(2\beta l)$	Open Stub Y/Y_0	Shorted Stub Y/Y_0
45°	$j0.42$	$-j2.42$
90°	$j1.0$	$-j1.0$
180°	∞	0
270°	$-j1.0$	$j1.0$

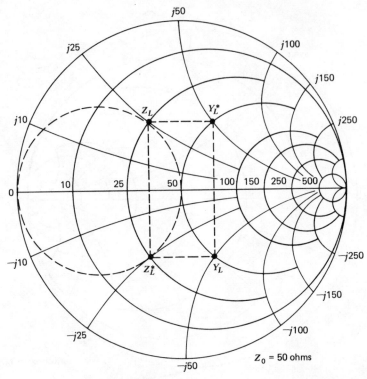

Figure 2.18 Smith Chart with $Z_L/Z_0 = 0.5 + j0.5$. Coordinates in ohms ($Z_0 = 50$ ohms).

center of the chart, the same result has been achieved (i.e., zero reflection coefficient and maximum power transfer).

The lossless impedance matching elements are summarized in Table 2.7. The impedance matching can always be done with only lumped LC, or only microstripline, or a combination of the two. A graphical solution on the Smith Chart usually gives the best insight into the design.

Table 2.7 Impedance Matching Elements

Matching Element	Lumped	Distributed Microstripline
Inductance series	$Z = j\omega L$	Cascade microstripline
Inductive shunt	$Y = \dfrac{1}{j\omega L}$	Parallel shorted stub
Capacitive series	$Z = \dfrac{1}{j\omega C}$	Not used
Capacitive shunt	$Y = j\omega C$	Parallel open stub

The impedance matching problem will be illustrated by matching Z_L/Z_0 $= 0.15 + j0.60$ to the center of the chart (i.e., $Z_L/Z_0 = 1$). First, the problem will be solved by lossless L sections, which can give four solutions. Next, the problem will be solved using microstripline sections, which can give four solutions. Next, a complete solution in both impedance and admittance planes with seventeen designs will be given to illustrate the wide variety of exact solutions that is possible.

Beginning with the lossless L sections in Fig. 2.19, we see that the first two solutions use a series capacitor followed by either a shunt capacitor (A) or a shunt inductor (B). The next solutions begin in the Y plane with a shunt capacitor followed by either a series capacitor (C) or a series inductor (D). For these solutions, part of the match is in the Z plane and part of the match is in the Y plane. The circle $R = 1$ and its mirror image are the primary curves required for the graphical solution.

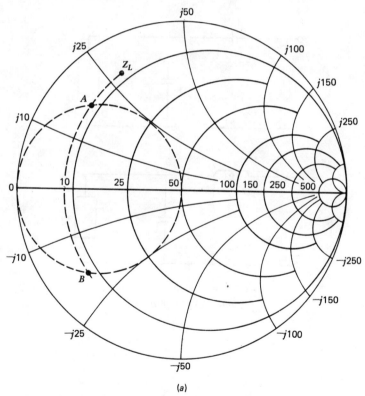

(a)

Figure 2.19 Impedance matching with lumped elements. (a) Coordinates in ohms ($Z_0 = 50$ ohms) ($Z_L/Z_0 = 0.15 + j0.60$). (b) Four matching circuits.

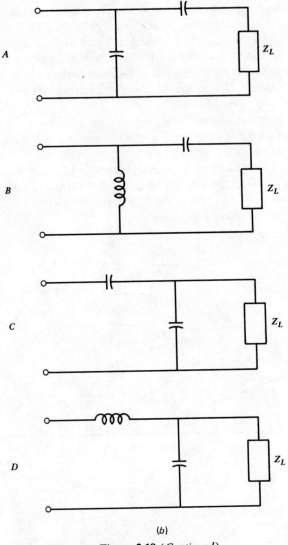

(b)

Figure 2.19 (*Continued*)

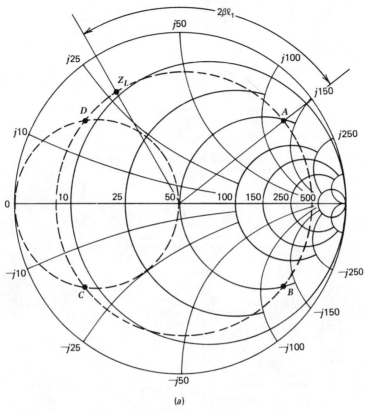

(a)

Figure 2.20 Impedance matching with a 50-ohm microstrip line element. (a) Coordinates in ohms ($Z_L/Z_0 = 0.15 + j0.60$). (b) Six matching circuits.

Now consider the solutions in Fig. 2.20, drawing first the effect of a cascaded 50-ohm line. When the line intersects point A in the Z plane, a series capacitor completes the match. At point B a series inductor is required. For point C a shunt inductor or shorted microstripline will give a match. At point D a shunt capacitor or open-circuited microstripline completes the impedance match.

The complete matching problem with multiple solutions is given in Fig. 2.21. First, a circle of constant $|\Gamma_L|$ is drawn to illustrate the effect of a cascaded 50-ohm line. A total length of $\lambda_g/2$ will give an input impedance of Z_L/Z_0 again. This represents one rotation around the chart. The impedance matching problem is visualized as moving from Z_L/Z_0 to the center of the chart, which is the generator. Clockwise rotation around the Smith Chart is required when transmission lines are cascaded. When the real axis

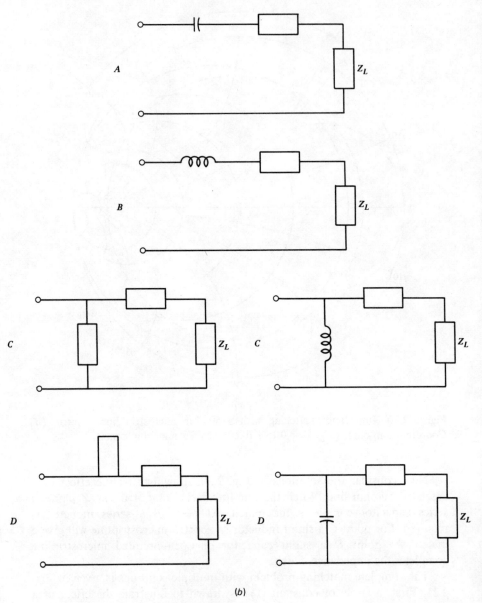

(b)

Figure 2.20 (*Continued*)

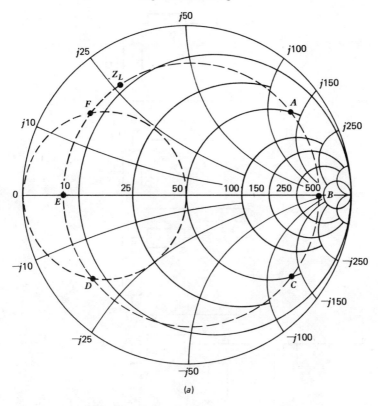

(a)

Figure 2.21 An impedance matching problem. (*a, b*) *Z* plane solutions; (*c, d*) *Y* plane solutions. [(*a, c*) coordinates in ohms ($Z_0 = 50$ ohms) ($Z_L/Z_0 = 0.15 + j0.60$).]

is located (points *B* and *E*), a $\lambda/4$ transmission line will always match to the center of the chart with

$$Z_T^2 = Z_1 Z_2 \tag{2.87}$$

with $Z_1 =$ point *B* or *E*
$\quad Z_2 =$ center of chart $= 50$ ohms

Notice that the mirror image of the circle $R = 1$ is useful for visualizing the L section networks using shunt admittance elements (see Fig. 2.21*a*). Likewise, in Fig. 2.21*c* the mirror image of the circle $G = 1$ is useful in plotting the shunt admittance values. Notice in Fig. 2.21*c* that the contour

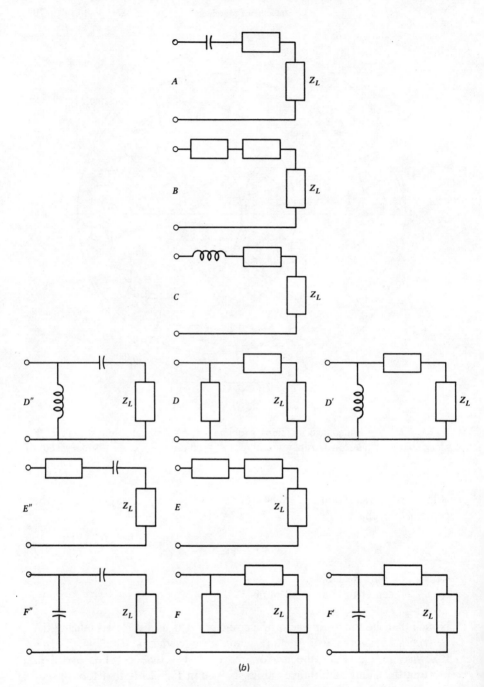

(b)

Figure 2.21 (*Continued*)

80

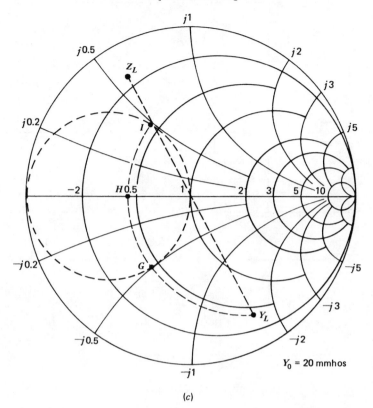

(c)

Figure 2.21 (*Continued*)

from Y_L is a constant admittance circle (not constant $|\Gamma_L|$), which is nearly the same contour for $|\Gamma_L| > 0.8$. Similarly, for the solution D'', E'', F'' in Fig. 2.21b, a constant reactance circle is used. A total of 17 solutions has been found using only two lossless reactive elements. These solutions are given in Table 2.8 for the reader to verify the impedance matching techniques.

Using only lumped elements, a total of four solutions have been found (D'', F'', G', I'). It can easily be shown that if the Z_L or Y_L point falls within the unit resistance or conductance circle, the number of lumped element L sections is only two instead of four. These solutions are indicated in Fig. 2.22.

Other impedance matching networks can be derived using the graphical techniques shown in Figs. 2.23 and 2.24. In the first technique, Z_L can be considered a point on the $\lambda/4$ matching section. The total length of the

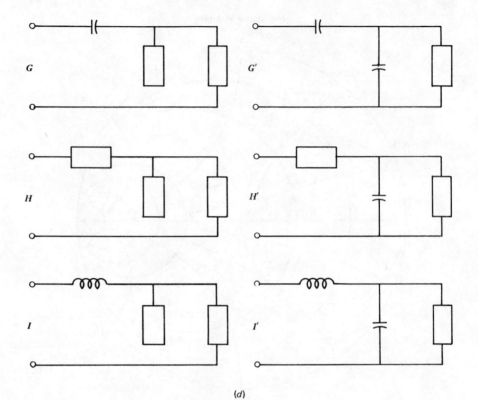

(d)

Figure 2.21 (*Continued*)

Table 2.8   Element Values for Solution of Fig. 2.21

	First Element	Second Element
A	$Z_0 = 50$ ohms, $l = 0.112\,\lambda$	$C = 0.30$ pf
B	$Z_0 = 50$ ohms, $l = 0.163\,\lambda$	$Z_0 = 154$ ohms, $l = 0.250\,\lambda$
C	$Z_0 = 50$ ohms, $l = 0.194\,\lambda$	$L = 5.2$ nH
D	$Z_0 = 50$ ohms, $l = 0.362\,\lambda$	$Z_0 = 50$ ohms, $l = 0.058\,\lambda$
D'	$Z_0 = 50$ ohms, $l = 0.362\,\lambda$	$L = 0.70$ nH
D''	$C = 0.90$ pF	$L = 0.70$ nH
E	$Z_0 = 50$ ohms, $l = 0.413\,\lambda$	$Z_0 = 16.6$ ohms, $l = 0.250\,\lambda$
E''	$C = 1.3$ pF	$Z_0 = 19.4$ ohms, $l = 0.250\,\lambda$
F	$Z_0 = 50$ ohms, $l = 0.444\,\lambda$	$Z_0 = 50$ ohms, $l = 0.192\,\lambda$
F'	$Z_0 = 50$ ohms, $l = 0.444\,\lambda$	$C = 2.0$ pF
F''	$C = 3.0$ pF	$C = 2.0$ pF
G	$Z_0 = 50$ ohms, $l = 0.129\,\lambda$	$C = 0.60$ pF
G'	$C = 0.90$ pF	$C = 0.60$ pF
H	$Z_0 = 50$ ohms, $l = 0.161\,\lambda$	$Z_0 = 81$ ohms, $l = 0.250\,\lambda$
H'	$C = 1.3$ pF	$Z_0 = 81$ ohms, $l = 0.250\,\lambda$
I	$Z_0 = 50$ ohms, $l = 0.181\,\lambda$	$L = 2.6$ nH
I'	$C = 1.7$ pF	$L = 2.6$ nH

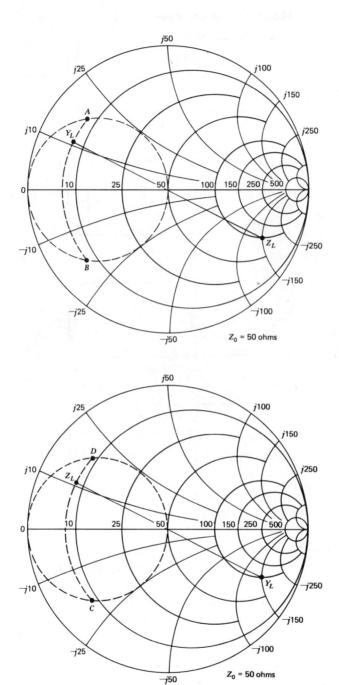

Figure 2.22 L section matching with Z_L (or Y_L) inside unit resistance (or conductance) circle. (a, b) coordinates in ohms. (c) Four matching circuits.

83

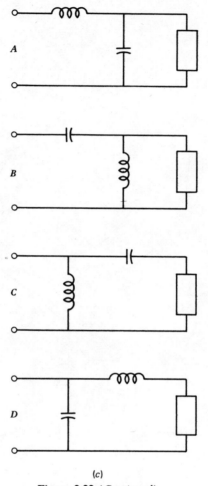

(c)

Figure 2.22 (*Continued*)

matching structure is less than $\lambda/4$, but the method only works if Z_L falls within the unit resistance or conductance circle. After connecting Z_L to the center of the chart, the perpendicular bisector is located on the real axis at point C. Draw a circle around C giving point r_1, which determines Z_T from (2.87).

$$\frac{Z_T}{Z_0} = \sqrt{r_1} \tag{2.88}$$

Now renormalize Z_L to Z_T. Move toward the generator until the real axis is

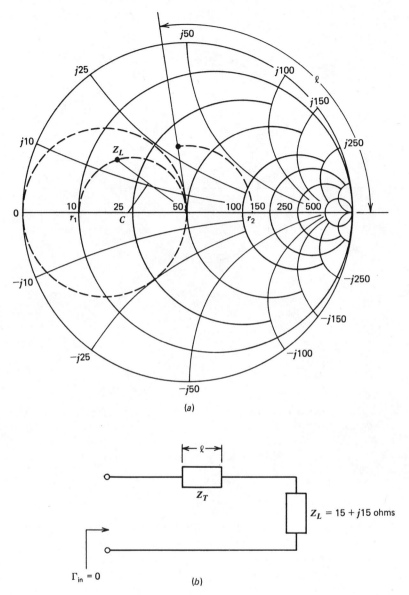

(a)

(b)

Figure 2.23 Single-section matching. (*a*) Coordinates in ohms. $Z_0 = 50$ ohms; $Z_L/Z_0 = 0.3 + j0.3$; $Z_T/Z_0 = \sqrt{0.2} = 0.45$; $Z_L'/Z_T = 0.67 + j0.67$; $r_2 = 2.22(50)$ ohms; $Z_{in}/Z_0 = 2.22(0.45) = 1.0$. Solution: $Z_T = 22.5$ ohms; $\beta l = 47^0$ (*b*) Matching circuit.

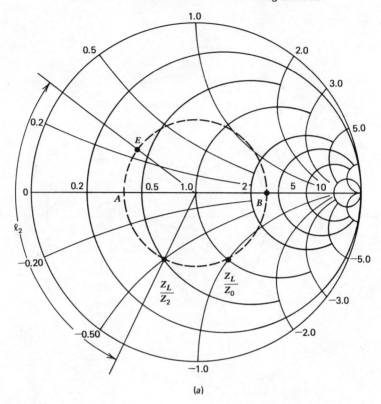

(a)

Figure 2.24 Two-section matching. (a) Normalized coordinates. $A = 40$ ohms; $B = 250$ ohms; $Z_2 = 100$ ohms; $E = 40 + j17.5$ ohms; (b) Normalized coordinates. $A' = 40$ ohms; $B' = 250$ ohms; $Z_1 = 25$ ohms; $D = Z_0/Z_1$; $E' = 40 + j17.5$ ohms. (c) $Z_L/Z_0 = 1 - j1$; $Z_0 = 50$ ohms. One solution: $l_2 = 0.119\,\lambda$, $l_1 = 0.048\lambda$, $Z_2 = 100$ ohms, $Z_1 = 25$ ohms.

intersected at r_2. The length of the matching section is given from the Smith Chart construction in Fig. 2.23. Renormalizing from Z_T to Z_0 finds the load transferred to the center of the chart.

A variation on this technique is the use of two sections of cascaded transmission line, as shown in Fig. 2.24. This transformation is possible only if the values of Z_1 and Z_2 are chosen properly. First, Z_L/Z_2 is drawn as a circle moving toward the generator (clockwise). Next, circle AB is redrawn normalized to Z_1 to give circle $A'B'$. Now the generator is normalized to Z_1 (point D) and the generator point is moved toward the load to give point E' and l_1. Transforming E' to E gives length l_2. This technique shows that an

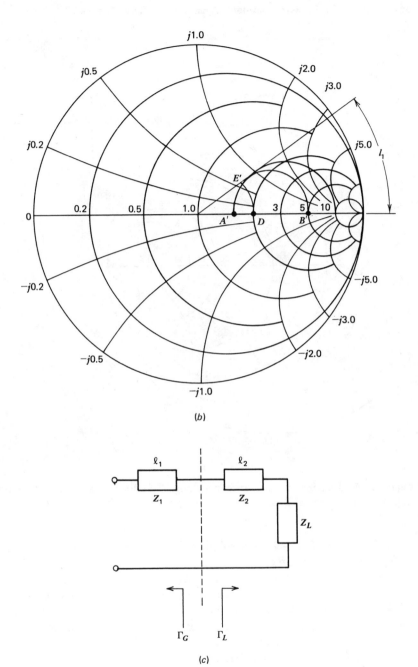

(b)

(c)

Figure 2.24 (*Continued*)

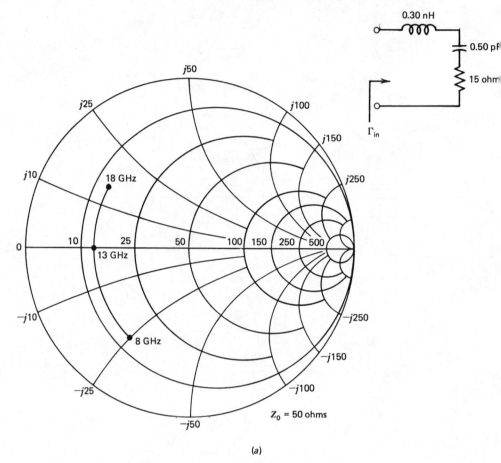

(a)

Figure 2.25 Broadband impedance matching. Coordinates in ohms. (a) Input impedance of GaAs MESFET. (b) Γ_1 plot. (c) Γ_2 plot.

infinite number of two-element solutions is always possible, since the range of Z_1 and Z_2 is very large.

The broadband impedance matching problem is considerably more difficult, since the load impedance is usually varying with frequency. The work of Bode and Fano has shown that there is a physical limitation on broadband matching when a reactive element is present. In the general case, Fano's limit gives a fundamental limitation on how well the matching can be achieved over a band of frequencies. For a high Q load with a large reactance change with frequency, a relatively poor match will be achieved; a low Q load can be matched much better over a broad bandwidth. The best

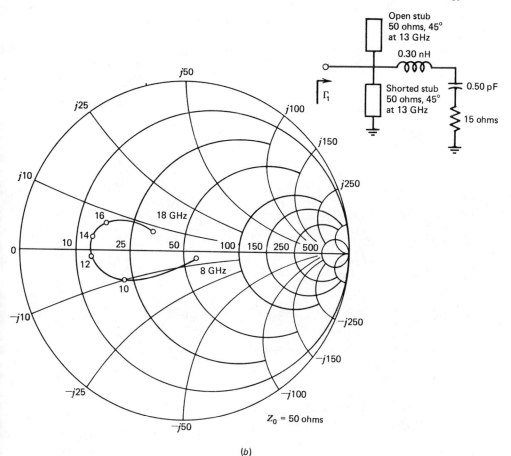

(b)

Figure 2.25 (*Continued*)

possible match for a parallel RC load of Q_1 over a bandwidth of Q_2 is given by

$$|\Gamma_{\min}| = e^{-\pi Q_2/Q_1} \tag{2.89}$$

where

$$Q_1 = R_1 C_1 \omega_0 \tag{2.90}$$

$$Q_2 = \frac{f_0}{\Delta f} = \frac{f_0}{BW} \tag{2.91}$$

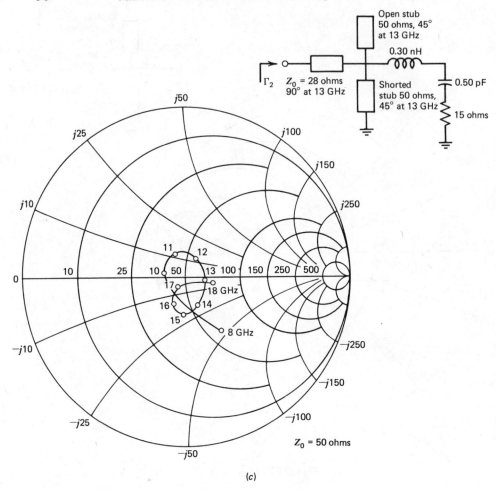

(c)

Figure 2.25 (*Continued*)

The technique for broadband matching consists of adding series-parallel-series resonant structures with the intent of approaching the $|\Gamma_{\min}|$ given by (2.89). As an example, consider the problem shown in Fig. 2.25, a simple model of the input circuit of a high-Q, low-loss GaAs MESFET. The solution is to series resonate the input, then parallel resonate, and so on until a minimum input reflection coefficient is realized. There are many circuits that can be used to realize these broadband matches, and more discussion of this problem will be presented later. (See Section 3.5.)

Bibliography

2.1 R. Allison, "Silicon Bipolar Microwave Power Transistors," *IEEE Transactions of MTT*, Vol. MTT-27, May 1979, pp. 415–422.

2.2 J. A. Archer, "Design and Performance of Small-Signal Microwave Transistors," *Solid State Electronics*, Vol. 15, March 1972, pp. 249–258.

2.3 J. A. Archer, "Low-Noise Implanted Base Microwave Transistors," *Solid State Electronics*, Vol. 17, April 1974, pp. 387–393.

2.4 E. Belohoubek and E. J. Denlinger, "Loss Considerations for Microstrip Resonators," *IEEE Transactions on MTT*, Vol. MTT-23, June 1975, pp. 522–526.

2.5 G. E. Brehm and G. D. Vendelin, "Biasing FETs for Optimum Performance," *Microwaves*, February 1974, pp. 39–44.

2.6 J. T. C. Chen and C. P. Snapp, "Bipolar Microwave Linear Power Transistor Design," *IEEE Transactions on MTT*, Vol. MTT-27, May 1979, pp. 423–430.

2.7 P. E. Day, "Transmission Line Transformation Between Arbitrary Impedance Using the Smith Chart," *IEEE Transactions on MTT*, Vol. MTT-23, September 1975, p. 772.

2.8 G. N. French and E. H. Fooks, "The Design of Stepped Transmission-Line Transformers," *IEEE Transactions on MTT*, Vol. MTT-16, October 1968, p. 885.

2.9 G. N. French and E. H. Fooks, "Double Section Matching Transformers," *IEEE Transactions on MTT*, Vol. MTT-17, September 1969, p. 719.

2.10 K. C. Gupta, R. Garg, and I. J. Bahl, *Microstrip Lines and Slotlines*, Artech House, Dedham, Mass., 1979.

2.11 T. H. Hsu and C. P. Snapp, "Low-Noise Microwave Bipolar Transistor with Sub-Half-Micrometer Emitter Width," *IEEE Transactions on Electron Devices*, Vol. ED-25, June 1978, pp. 723–730.

2.12 E. O. Johnson, "Physical Limitations on Frequency and Power Parameters of Transistors," *RCA Review*, Vol. 26, June 1965, pp. 163–177.

2.13 J. S. Lamming, "Microwave Transistors," in M. J. Howes and D. V. Morgan, Eds., *Microwave Devices*, Wiley, New York, 1976.

2.14 C. A. Liechti, "Microwave Field-Effect Transistors—1976," *IEEE Transactions on MTT*, Vol. MTT-24, June 1976, pp. 279–300.

2.15 G. I. Matthaei, "Short-Step Chebyshev Impedance Transformers," *IEEE Transactions on MTT*, Vol. MTT-14, August 1966, p. 372.

2.16 T. A. Milligan, "Transmission-Line Transformation Between Arbitrary Impedances," *IEEE Transactions on MTT*, Vol. MTT-24, March 1976, p. 159.

2.17 R. A. Pucel, D. J. Masse, and C. P. Hartwig, "Losses in Microstrip," *IEEE Transactions on MTT*, Vol. MTT-16, June 1968, pp. 342–350; and correcting pp. *IEEE Transactions on MTT*, Vol. MTT-16, December 1968, p. 1064.

2.18 M. V. Schneider, "Microstrip Dispersion," *Proceedings of the IEEE*, Vol. 60, January 1972, pp. 144–146.

2.19 P. H. Smith, *Electronic Applications of the Smith Chart*, McGraw-Hill, New York, 1969.

2.20 C. P. Snapp, "Bipolars Quietly Dominate," *Microwave System News*, November 1979, pp. 45–67.

2.21 P. I. Somlo, "A Logarithmic Transmission Line Chart," *IEEE Transactions on MTT*, Vol. MTT-10, July 1960, p. 463.

2.22 S. M. Sze, *Physics of Semiconductor Devices*, Wiley, New York, 1969.

2.23   G. D. Vendelin, "Five Basic Bias Designs for GaAs FET Amplifiers," *Microwaves*, February 1978, pp. 40–42.

2.24   G. D. Vendelin and M. Omori, "A Computer Model of the GaAs MESFET Valid to 12 GHz," *Electronics Letters*, Vol. 11, February 1975, pp. 60–61.

2.25   G. D. Vendelin and M. Omori, "Try CAD for Accurate GaAs MESFET Models," *Microwaves*, June 1975, pp. 58–70.

2.26   G. D. Vendelin, W. Alexander, and D. Mock, "Computer Analyzes RF Circuits with Generalized Smith Charts," *Electronics*, March 1974, pp. 102–109.

2.27   G. D. Vendelin, "Limitations on Stripline Q," *Microwave Journal*, May 1970, pp. 63–69.

2.28   G. D. Vendelin, "High-Dielectric Substrates for Microwave Hybrid Integrated Circuitry," *IEEE Transactions on MTT*, Vol. MTT-15, December 1967, pp. 750–752.

CHAPTER THREE

AMPLIFIER DESIGNS

3.0 Introduction

Microwave amplifier designs are usually completed using S-parameters, since this is the most accurate two-port description of the transistor. The impedance matching problem for various amplifier designs is discussed in this chapter. The simplicity of a one-stage design is first presented. The advantages of a balanced design are described. Next, the special design considerations of low noise, high power, and wide bandwidth are described. Since some designs benefit from feedback, this subject is briefly introduced. Finally, the design of two-stage amplifiers is also described in this chapter.

3.1 One-Stage Design

As we recall from Fig. 1.1, the design of a one-stage amplifier consists of finding:

1 An input lossless matching network M_1.
2 An output lossless matching network M_2.

so that the maximum or desired transistor gain is achieved over the operating bandwidth of the amplifier. Usually, the common-emitter or common-source configuration is chosen for highest gain per stage. If the stability factor k is greater than unity, these two networks can be found to give the maximum available gain G_{ma} given by (1.161). If the stability factor is less than or equal to unity, the amplifier could be terminated in a matching structure which causes oscillation, that is, G_{ma} is infinite. This should be avoided by locating the regions of instability in the Γ_G and Γ_L planes. The input and output terminations (Γ_G and Γ_L) must be designed to avoid the instability regions. Usually these unstable regions are near the conjugate match for S_{11} and S_{22}. Thus a stable amplifier will require some input and output mismatch if k is less than or equal to unity.

There are at least two alternate approaches for potentially unstable amplifiers:

1   Add resistive matching elements to make $k \geq 1$ and $G_{ma} = G_{ms}$.
2   Add feedback to make $k \geq 1$ and $G_{ma} = G_{ms}$.

In practice, neither of these approaches is recommended, since other properties of the amplifier are usually degraded: gain, bandwidth, noise figure, rf output power, and dc bias power. It is usually recommended to accept a transistor with $k < 1$, design the amplifier for a gain approaching G_{ms}, and to insure that the Γ_G and Γ_L terminations provide stability at all frequencies, both inside and outside the amplifier passband. The regions of instability for the output plane (the Γ_L plane) are given in Fig. 1.10, which was plotted from the transistor S-parameters and from (1.97) and (1.98). Similar regions can be found for the input plane (Γ_G plane).

The design of an amplifier would usually have the following specifications:

Gain and gain flatness.
Bandwidth and center frequency ($f_2 - f_1, f_0$).
Noise figure.
Linear output power.
Input reflection coefficient (VSWR).
Output reflection coefficient (VSWR).
Bias voltage and current.

For small-signal amplifiers, the small-signal S-parameters are sufficient to complete the design. After selecting an appropriate transistor based on these specifications, a one-stage amplifier design should be considered if sufficient gain can be achieved.

The circuit topology should be chosen in order to allow dc bias for the transistor. Usually, short-circuited stubs are placed near the transistor to allow dc biasing. If the topology does not allow dc bias, a broadband bias choke or high-resistance bias circuit that does not affect the amplifier performance must be used.

The following steps can be tabulated for the design of a one-stage amplifier:

1   Select transistor based upon S-parameters, noise figure, and linear output power.

2 Calculate k and G_{ma} or G_{ms} versus frequency.

3 For $k > 1$, select the topologies that match the input and output (and allow dc biasing) at the upper band edge f_2. This will give G_{ma} and $S'_{11} = S'_{22} = 0$ ideally. Usually, $S_{12} = 0$ is assumed for the initial design. Next, the topology may be varied to flatten gain versus frequency at the expense of S'_{11} and S'_{22}.

4 For $k < 1$, plot the regions of instability on the Γ_G and Γ_L planes and select topologies which partially match the input and output at the upper band edge and avoid the unstable regions. The gain will approach G_{ms} as an upper limit. Next, the topology may be varied to flatten gain versus frequency.

5 After finding initial M_1 and M_2, plot the amplifier S-parameters versus frequency; make adjustments in topology until the specifications for gain, input reflection coefficient, and output reflection are satisfied. Also plot Γ_G and Γ_L versus frequency to verify amplifier stability. Obviously, a computer can be very helpful with these calculations.

6 Design dc bias circuit. Lay out the elements of the complete amplifier and check realizability.

An example will illustrate the procedure. Consider the design of a single-stage amplifier with the following specifications:

Gain: 15 dB min (gain flatness ± 0.5 dB).
Frequency range: 3.0 to 4.0 GHz.
Input reflection coefficient: < 0.5.
Output reflection coefficient: < 0.5.
Bias voltage: 15 V.
Bias current: 100 mA maximum.

A GaAs MESFET is required if a single-stage design is to be adequate. The HFET 1001 is selected on the basis of the 4-GHz gain of 18 dB (G_{ma}).

The S-parameters of this device are given in Table 3.1, including the effect of chip bonding parasitics, which were

$$L_G = 0.30 \text{ nH}$$

$$L_S = 0.15 \text{ nH}$$

$$L_D = 0.30 \text{ nH}$$

Most transistor data sheets include the bonding parasitics, but Hewlett

Table 3.1 S-Parameters of HFET-1001 Chip with Bonding Inductances at $V_{DS} = 4.0$, $I_{DS} = I_{DSS}$

f (GHz)	S_{11}	S_{21}	S_{12}	S_{22}	k	G_{ma}/G_{ms} (dB)
1	0.95 ∠$-20°$	3.55 ∠$161°$	0.013 ∠$80°$	0.82 ∠$-5°$	0.43	24.4
2	0.90 ∠$-42°$	3.38 ∠$143°$	0.019 ∠$79°$	0.79 ∠$-8°$	0.59	22.5
3	0.81 ∠$-65°$	3.23 ∠$126°$	0.027 ∠$77°$	0.75 ∠$-15°$	0.78	20.8
4	0.71 ∠$-86°$	2.86 ∠$110°$	0.031 ∠$79°$	0.75 ∠$-17°$	1.06	18.2
5	0.63 ∠$-108°$	2.57 ∠$94°$	0.036 ∠$89°$	0.73 ∠$-21°$	1.14	15.6
10	0.57 ∠$+178°$	1.60 ∠$41°$	0.129 ∠$108°$	0.72 ∠$-56°$	0.52	10.9

Table 3.2 S-Parameters of Amplifier in Fig. 3.1b

f (GHz)	S_{11}	S_{21}	S_{12}	S_{22}	k	G_{ma}/G_{ms}	S_{21} (dB)
3	0.81 ∠$53°$	6.22 ∠$179°$	0.052 ∠$130°$	0.55 ∠$19°$	0.77	20.8	15.88
4	0.426 ∠$-128°$	6.10 ∠$69.5°$	0.066 ∠$38.5°$	0.26 ∠$159°$	1.07	18.2	15.70

Packard chip data sheets usually do not. This device is potentially unstable over the frequency range 0 to 3.8 GHz and above 7 GHz. An initial topology that allows dc bias is given in Fig. 3.1*a*. When calculated on the computer [or by means of (1.136)], this circuit gave a transducer gain of 16 dB at 4 GHz, but only 7 dB at 3 GHz. Next the computer was used to vary Z_0 and the lengths until the solution in Fig. 3.1*b* was found.

The resulting S-parameters for the 16-dB design are given in Table 3.2. Notice that the S_{11} and S_{22} values are larger than 0.50 at 3 GHz, but the gain is flat. This always occurs in a single-stage amplifier if the gain is

Initial Design

Element	Z_0 (ohms)	βl at 4 GHz
1	50	24.5°
2	50	26.5°
3	50	60.5°
4	50	24.5°

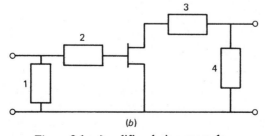

(a)

Final Design

Element	Z_0 (ohms)	βl at 4 GHz
1	100	23.7°
2	100	14.1°
3	100	57.8°
4	100	37.0°

(b)

Figure 3.1 Amplifier design example.

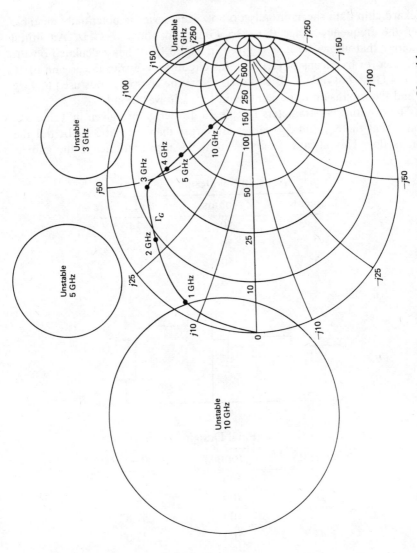

3.2 Unstable region in input plane and output plane. Coordinates in ohms. (*a*) Unstable region in input plane. (*b*) Unstable region in output plane.

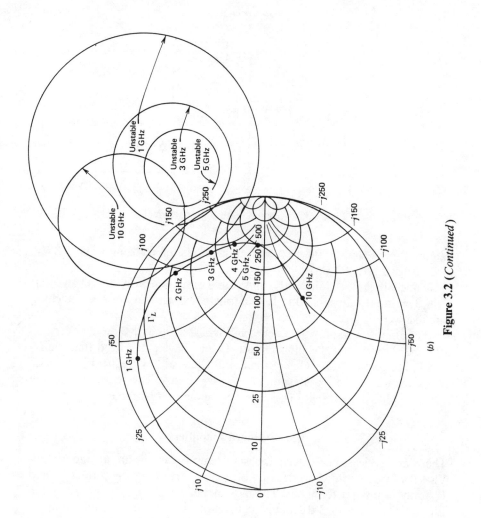

Figure 3.2 (*Continued*)

99

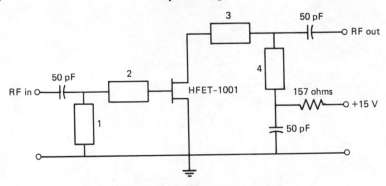

Figure 3.3   Complete amplifier schematic.

required to be flat. In the next section methods for realizing a flat VSWR will be discussed.

Notice that it is inherently impossible simultaneously to achieve flat gain, low input reflection, and low output reflection with lossless matching circuits.

The amplifier stability is checked in Fig. 3.2 for the final design over 1 to 10 GHz. Notice that the short circuited stubs have guaranteed that the Γ_G and Γ_L are on the "safe side" of the Smith Chart at low frequencies. The dc bias circuit is found from estimating $I_{\text{DSS}} \simeq 70$ mA. The schematic for the entire amplifier is given in Fig. 3.3. All elements are realizable.

3.2   Balanced Amplifiers

There are at least three techniques for improving the input and output VSWR of the amplifier while simultaneously presenting the correct Γ_G and Γ_L to the transistor terminals. The most popular technique is that shown in Fig. 3.4a, the use of 3-dB Lange couplers in a balanced configuration. In this technique the reflections from the identical amplifiers all appear at the termination port, so that the input port appears matched.

One can show that

$$S_{11} = \tfrac{1}{2}(S_{11A} - S_{11B}) \tag{3.1}$$

The operating bandwidth of the amplifier is limited by the bandwidth of the coupler, which can be made very wide (more than 2 octaves). This technique of building one-stage balanced amplifiers has many advantages:

1   The transistor can be intentionally mismatched for gain flatness, noise figure, output power, stability, and so on.

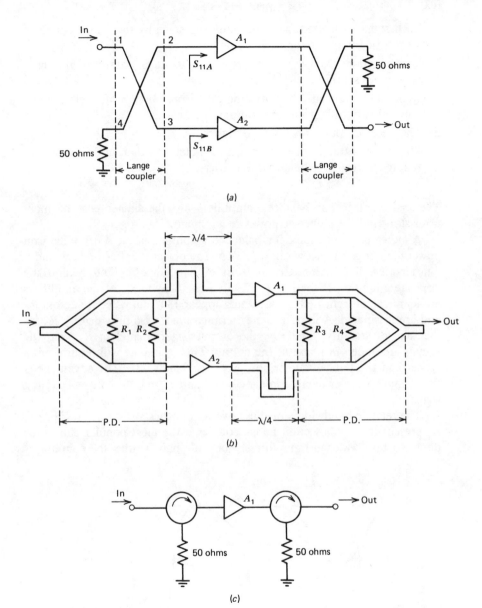

Figure 3.4 Low-VSWR amplifiers, (*a*) Balanced amplifier using 3-dB Lange coupler; (*b*) Balanced amplifier using 3-dB Wilkinson power divider (P.D.); (*c*) Isolator amplifier.

101

2 Each stage is isolated from the following stage by the Lange couplers. Very little interstage tuning is needed.

3 The reflections from A_1 and A_2 are terminated in 50 ohms, which usually guarantees stability.

4 If one transistor fails, the operating gain drops about 6 dB, which can be a useful feature in some applications.

5 The linear output power is 3 dB higher.

6 The labor in tuning these amplifiers is much less than in designing and tuning multistage unbalanced amplifiers.

The disadvantages of balanced amplifiers are the higher cost of more transistors and the higher dc power requirement.

Another popular choice for balanced amplifiers is the 3-dB Wilkinson power divider circuit shown in Fig. 3.4b. The power is split in phase, and a quarter-wave line is inserted in front of one amplifier and behind the opposite amplifier. If the input reflection coefficients S'_{11} of the amplifiers are identical, all reflected input signals appear 180° out of phase across R_2 and are dissipated. A low-frequency termination is also provided by R_2 for the sake of stability. Similarly, the output is terminated in R_3, and the signals are added in phase at the output. This technique will require more space than the Lange coupler and gives less bandwidth. The advantage is that a much simpler microstripline circuit can be used (no narrow coupling strips).

The relative bandwidths of these two approaches are shown in Fig. 3.5. At present the single-section Lange coupler is the most popular choice for double-octave wideband amplifiers. For wide bandwidths, the coupling is

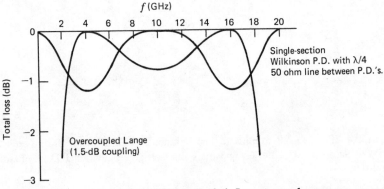

Figure 3.5 Comparison of 10-GHz cascaded Lange couplers versus cascaded single-section Wilkinson power dividers.

tighter than 3 dB (about 1.5 dB seems optimum). The third approach to low-VSWR amplifiers is the use of broadband isolators. This approach requires a thin-film ferrite circuit and a magnetic structure for realizing the isolators. Moreover, low-frequency broadband isolators are not available below 2 GHz. Unless a proven isolator capability exists, this approach is not commonly used.

In summary, the most common amplifier designs use the 3-dB Lange coupler to achieve low VSWR over broad bandwidths. These balanced amplifiers are easily cascaded to achieve high gain and high output powers.

3.3 Low-Noise Design

For the low-noise design the transistor data must include the S-parameters at the low-noise bias and four noise parameters. One common noise parameter set is

$$
\begin{aligned}
&F_{\min} &&\text{Minimum noise figure}\\
&R_n = N/G_G &&\text{Noise resistance}\\
&Y_{on} = G_{on} + jB_{on} &&\text{Optimum noise admittance}
\end{aligned}
$$

The noise figure of the two-port is given by the source admittance (or impedance) presented to the input terminals and is calculated from (1.182), which is repeated here:

$$ F = F_{\min} + \frac{R_n}{G_G} |Y_G - Y_{on}|^2 $$

The output port is tuned for the maximum available gain if the amplifier is single-stage. In a two-stage, low-noise design the interstage circuit will probably be tuned for minimum second-stage noise figure. The noise figure of a multistage amplifier is given by

$$ F_{\text{tot}} = F_1 + \frac{F_2 - 1}{G_{\text{av}1}} + \frac{F_3 - 1}{G_{\text{av}1}G_{\text{av}2}} + \cdots \tag{3.2} $$

where $G_{\text{av}1}$ = available gain of first stage
$G_{\text{av}2}$ = available gain of second stage

If all stages are designed for minimum noise figure, we find

$$ (F_{\text{tot}})_{\min} = (F_{\min} - 1) + \frac{F_{\min} - 1}{G_{\text{av}}} + \frac{F_{\min} - 1}{(G_{\text{av}})^2} + \cdots + 1 \tag{3.3} $$

Using

$$\frac{1}{1-x} = 1 + x + x^2 + \cdots \tag{3.4}$$

we find a quantity $(F_{tot} - 1)$ which is defined as noise measure M. The minimum noise measure

$$(F_{tot})_{min} - 1 = \frac{F_{min} - 1}{1 - 1/G_{av}} = M_{min} \tag{3.5}$$

refers to the noise of an infinite chain of optimum tuned, low-noise stages, so it represents a lower limit on the noise of the amplifier. Since it differs from $(F_{tot})_{min}$ by a factor of -1, the noise measure can be less than unity, but the noise figure cannot be less than unity. The noise performance of a transistor is often visualized by plotting noise circles on the Γ_G plane (see Fig. 3.6). This technique allows the designer to visualize the effect of nonoptimum tuning so that practical noise performance can be estimated.

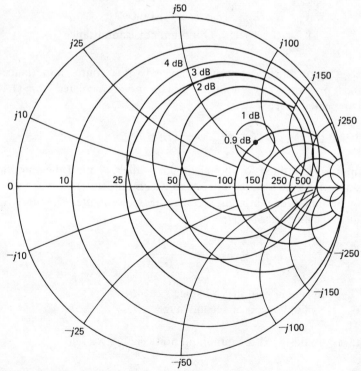

Figure 3.6 Noise figure circles for microwave transistor in Γ_G Plane. Coordinates in ohms. Dexcel 2503B chip. $V_{DS} = 3.5$ V; $I_{DS} = 15$ mA; $f = 4$ GHz; $F_{min} = 0.9$ dB; $\Gamma_{on} = 0.53 \angle 30°$; $R_n/50 = 0.97$.

Some additional effects on low-noise performance are 3-dB Lange coupler losses, microstripline losses, nonoptimum dc biasing, and resistive losses. Feedback can also be used in low-noise designs to vary F_{min} and Y_{on}. When feedback is used, the noise figure may be reduced, but the available gain is also reduced. The minimum noise measure is invariant to lossless feedback elements. It is not invariant to the common lead and is usually lower for common-emitter or common-source transistors.

3.4 High-Power Design

For high-power amplifier design, the large-signal parameters of the transistor should be obtained. For Class A operation (see Fig. 2.9), the small-signal S-parameters are a rough approximation of the large-signal performance. For Class AB, B, or C the small-signal S-parameters become progressively less useful.

The large-signal performance of the transistors should be measured under the maximum allowed dc bias. The optimum large-signal generator and load impedances (Γ_{GP} and Γ_{LP}) must be measured at each frequency of operation. These parameters will resemble S_{11}^* and S_{22}^*, but usually the output Γ_{LP} is significantly changed from S_{22}^*. The gain is reduced in comparison to G_{ma} or G_{ms}.

The contours of constant output power can be plotted in the Γ_L and Γ_G planes to determine the effect of nonoptimum match. This is analogous to noise circles in the Γ_G plane, but the contours are not necessarily circles. By assuming circles, we may write the effect of output load mismatch on large-signal gain, G_{LS}.

$$G_{LS} = G_{max} - \frac{K_1}{G_L} |Y_L - Y_{L\,opt}|^2 \qquad (3.6)$$

There are four large-signal parameters of importance in the output plane:

G_{max} the maximum large signal gain
K_1 an empirical factor
$Y_{LOPT} = G_{LOPT} + j\,B_{LOPT}$, the optimum load admittance

The input matching Γ_{GP} and the operating bias point would complete the data needed for large-signal design. If a wideband high-power amplifier is required, the input match must be designed to flatten the gain while the output match is designed to give maximum power, Γ_{LP}.

An example of large-signal detuning is plotted in Fig. 3.7 for the Dexcel 3501A chip at 12 GHz. Contours of Γ_G detuning and Γ_L detuning are both

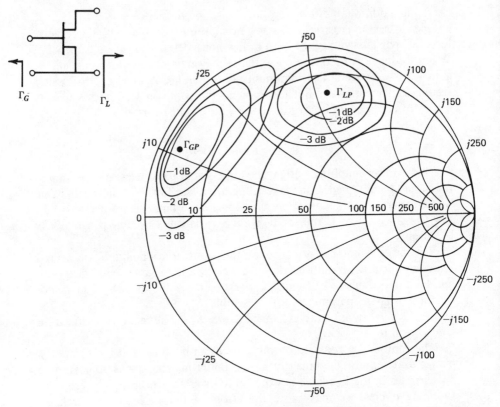

Figure 3.7 Large-signal tuning of GaAs MESFET. Coordinates in ohms. Dexcel 3501A chip. $V_{DS} = 8.0$ V; $I_{DS} = 0.5 I_{DSS}$; $f = 12$GHz. Optimum tuned for $P_{out} = +18.5$ dBm at 1 dB gain compression. Gain $= 6.5$ dB.

plotted. The value of Γ_{GP} is very close to S_{11}^*, but Γ_{LP} has changed significantly from S_{22}^* at 12 GHz.

Another important amplifier specification is the dynamic range, which is the range of input or output power with linear gain. At low powers this is limited by the noise figure or the minimum detectable signal, and at high powers it is limited by the power level where small-signal gain has been compressed by 1 dB. This is indicated schematically in Fig. 3.8. Defining the minimum detectable signal as 3 dB above thermal noise, we have

$$(\text{MDS})_{out} = kTB + 3 \text{ dB} + NF + G \qquad (3.7)$$

$$(\text{MDS})_{in} = kTB + 3 \text{ dB} + NF \qquad (3.8)$$

$$kTB = -114 \text{ dBm/MHz} \qquad (3.9)$$

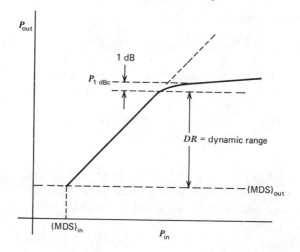

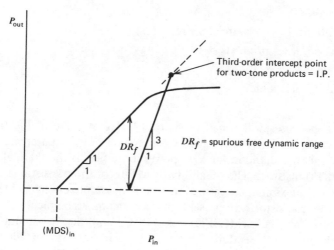

Figure 3.8 Dynamic range of amplifiers.

The dynamic range becomes:

$$DR = P_{1\ dBc} - (MDS)_{out} \tag{3.10}$$

When two signals close in frequency are present in the amplifier, a third-order product will appear in the passband at $2f_2 - f_1$ or $2f_1 - f_2$ as a result of the nonlinear components of g_m. Since the intercept point of the third-order two-tone products is usually 10 dB above the 1-dB gain compression point, the spurious free dynamic range can be calculated graphically. The spurious free dynamic range is given by

$$DR_f = \tfrac{2}{3}\left[\text{I.P.} - G - (\text{MDS})_{\text{in}}\right] \qquad (3.11)$$

For an amplifier with a bandwidth of 30 MHz, transducer gain of 30 dB, and noise figure (NF) of 6 dB, the input minimum detectable signal is

$$(\text{MDS})_{\text{in}} = -114 \text{ dBm} + 15 \text{ dB} + 6 \text{ dB} + 3 \text{ dB}$$

$$= -90 \text{ dBm}$$

If the 1-dB compression point is at $+15$ dBm, the linear dynamic range is 75 dB, and the spurious free dynamic range is about 57 dB.

High-power designs can also be optimized for any of the following:

1 High power ($\Gamma_L = \Gamma_{LP}$).
2 High gain ($\Gamma_L \simeq S_{22}^*$).
3 High efficiency (no source bias resistors).
4 Low third-order products.
5 Low second harmonics.
6 Low third harmonics.

Each of these cases will usually require a different value of Γ_L, the load tuning parameter. There is a tradeoff between all of these parameters.

The final consideration for high-power amplifiers is the thermal impedance of the transistors. The reliability or MTBF of the transistor is directly related to the maximum operating junction temperature or channel temperature of the transistor. At present the maximum recommended channel temperature for GaAs MESFETs is 175°C, and the maximum recommended junction temperature for the Si bipolar transistor is 200°C. The higher number for Si is a result of the maturity of this technology. On the basis of the wider bandgap of GaAs, a higher operating channel temperature for GaAs should become possible, but at present it is limited to 175°C by a potential chemical reaction at higher temperatures. For a medium-power GaAs MESFET on a conventional 0.005-inch thick chip, the thermal resistance of the chip is about 75°C/W for a 500 μm $\times$ 1 μm gate. When operating at 0.5 I_{DSS}, the dc power dissipation is

$$I_{\text{DSS}} = 0.12 \text{ amp}$$
$$P_{\text{diss}} = \text{VI} = 10(.5)(.12) = .60 \text{ W}$$
$$\Delta T = 45°\text{C}$$
$$T_c = T_A + \Delta T = 70°\text{C}$$

However, when operating at an ambient temperature of 90°C, the channel temperature is 135°C, approaching the maximum allowable value of 175°C.

3.5 Broadband Design

The broadband amplifier design problem is solved by considering the power gain rolloff of the transistor with frequency (usually 6 dB/octave), the input equivalent circuit, the output equivalent circuit, the gain bandwidth limitations of the input, the gain bandwidth limitations of the output, and the overall amplifier stability versus frequency. The general design problem of a one-stage broadband amplifier is outlined schematically in Fig. 3.9. The gain of the matching network may be sloped in the input circuit M_1, the output circuit M_2, or both, depending on the gain-bandwidth limitation of the device.

The gain of the transistor versus frequency is given by

$$G_A(f) = G_0\left(\frac{f}{f_2}\right)^{-k}$$ (3.12)

where G_0 is the gain at f_2,

$$k = \frac{x}{10\log 2} \simeq \frac{x}{3}$$ (3.13)

and x is the slope in dB/octave. Usually x is 6 dB/octave and $k = 2$ (not to be confused with the stability factor k). Since the overall gain is intended to be constant from f_1 to f_2 at G_0, the matching circuit should be

$$G_M(f) = K\left(\frac{f}{f_2}\right)^{k} \quad f_1 < f < f_2$$ (3.14)

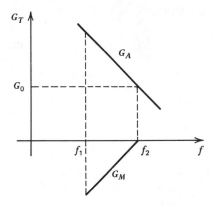

Figure 3.9 Gain versus frequency.

The parameter K is the gain scale factor, which is less than or equal to unity. The overall amplifier gain is

$$G_T = G_M(f)G_A(f)$$
$$= G_{M1}(f)G_{M2}(f)G_A(f)$$
$$= KG_0 \tag{3.15}$$

where G_{M1} is associated with the input matching network and G_{M2} is associated with the output matching network. These parameters are chosen from considerations of the gain-bandwidth limitations.

Over a broad bandwidth, the input and output impedances of a transistor can usually be represented by a resistor and reactance model. The four general cases are given in Fig. 3.10 along with the corresponding expression for Fano's limit. Two of these cases are high-pass and two are low-pass with frequency. The GaAs MESFET chip can usually be represented by Fig. 3.10c for the input (high-pass) and Fig. 3.10a for the output (low-pass). The Si bipolar transistor chip can usually be represented by Fig. 3.10b for the input (low-pass) and by Fig. 3.10b for the output (low-pass). Each case of transistor S-parameter data should be individually examined for the appropriate reactive model. Obviously, right-hand-plane zeros of ρ in the complex frequency plane or s plane are avoided for maximum bandwidth.

In the matching circuit, the parasitic element of the transistor must be the first element of the matching circuit. Since there is no control over this reactive element and the transistor resistance, this imposes a limit on the bandwidth where a good match can be obtained. It will turn out that even greater bandwidth can be achieved by including a gain slope in the matching circuit as shown in Fig. 3.9, but this will be shown later.

The details of broadband matching for a GaAs MESFET will be considered below (see Fig. 3.11). The input is usually the high-pass circuit of Fig. 3.10c and the output is the low pass circuit of Fig. 3.10a. For the network M_1 and C_{in}, which is lossless and reciprocal,

$$|\Gamma_1|^2 = 1 - |S_{12}|^2_{in} = 1 - G_{M1}(f) \tag{3.16}$$

$$|\Gamma_1| = |\rho_{in}| \tag{3.17}$$

$$\int_0^\infty \frac{1}{\omega^2} \ln\left|\frac{1}{\rho_{in}}\right| d\omega \le \pi R_{in} C_{in} \tag{3.18}$$

Normalizing frequency to $\omega_2 = 1$ gives

$$\Omega = \frac{\omega}{\omega_2} \tag{3.19}$$

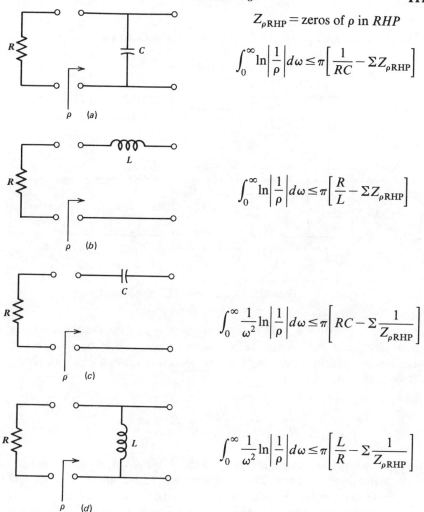

$$Z_{\rho\text{RHP}} = \text{zeros of } \rho \text{ in } RHP$$

$$\int_0^\infty \ln\left|\frac{1}{\rho}\right| d\omega \leq \pi\left[\frac{1}{RC} - \Sigma Z_{\rho\text{RHP}}\right]$$

$$\int_0^\infty \ln\left|\frac{1}{\rho}\right| d\omega \leq \pi\left[\frac{R}{L} - \Sigma Z_{\rho\text{RHP}}\right]$$

$$\int_0^\infty \frac{1}{\omega^2} \ln\left|\frac{1}{\rho}\right| d\omega \leq \pi\left[RC - \Sigma \frac{1}{Z_{\rho\text{RHP}}}\right]$$

$$\int_0^\infty \frac{1}{\omega^2} \ln\left|\frac{1}{\rho}\right| d\omega \leq \pi\left[\frac{L}{R} - \Sigma \frac{1}{Z_{\rho\text{RHP}}}\right]$$

Figure 3.10 Broadband impedance match limit, or Fano's limit.

which makes (3.18) become

$$\int_0^\infty \frac{1}{\Omega^2} \ln\left|\frac{1}{\rho_{\text{in}}}\right| d\Omega \leq \pi\omega_2 R_{\text{in}}C_{\text{in}} = \pi\tau_{HP} = \int_{\omega_1}^1 \frac{1}{\Omega^2} \ln\left|\frac{1}{\rho_{\text{in}}}\right| d\Omega \qquad (3.20)$$

$$\tau_{HP} = \omega_2 R_{\text{in}}C_{\text{in}} \qquad (3.21)$$

Thus ρ_{in} cannot be zero over a finite bandwidth; moreover, the wider the

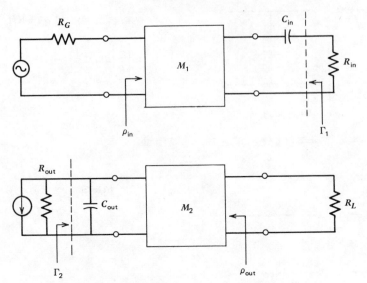

Figure 3.11 Broadband GaAs MESFET amplifier design.

bandwidth, the higher ρ_{in} will become. By sloping ρ_{in} over the band, a lower value of ρ_{in} can be obtained at ω_2 and a higher value near ω_1, where mismatch is usually required. For the sloped matching network, using (3.14), (3.16), and (3.17) gives

$$\int_{\omega_1}^{1} \frac{1}{\Omega^2} \ln \left| \frac{1}{1 - K(f/f_2)^k} \right| d\Omega < 2\pi\tau_{HP} \qquad (3.22)$$

The goal in wideband amplifier design is to achieve $K = 1$, no gain reduction at the high end of the band. The integration of (3.22) is plotted in Fig. 3.12 to indicate the gain-bandwidth limitations of the input.

For no gain reduction at f_2, the value of τ_{HP} must be as large as possible. If a 6-dB/octave slope is allowed at the input, the value of τ_{HP} may be slightly lower. If the value of τ_{HP} is even smaller, either the bandwidth must be reduced or the matching circuit must introduce some mismatch loss at the high-frequency end of the band.

An example will illustrate the use of these curves. Consider the broadband design of a GaAs MESFET amplifier with

$$R_{in} = 10 \text{ ohms}$$

$$C_{in} = 0.5 \text{ pF}$$

$$f_2 = 18 \text{ GHz}$$

The designer is asked for designs of

1 9 to 18 GHz
2 6 to 18 GHz
3 1.8 to 18 GHz

If a 6-dB/octave gain rolloff is allowed to compensate the transistor gain rolloff completely, the result is tabulated in Table 3.3. If 3 dB/octave is allowed to partially compensate the transistor gain rolloff, the result is also given in Table 3.3.

If the design bandwidth is in a lower frequency range, less bandwidth is possible (but higher gain). Consider the results for

4 1 to 2 GHz
5 1 to 3 GHz
6 1 to 10 GHz

The gain reduction for these bandwidths is also tabulated in Table 3.3.

Thus, for wideband amplifiers, no gain reduction is required if the upper frequency is high enough or the gain slope is high enough to allow Fano's integral relations to be satisfied. An acceptable wideband FET must have a large gain at f_2 and a large high-pass time constant.

Another example of the usefulness of the input bandwidth restriction in Fig. 3.12 is the maximum usable bandwidth from this transistor. Consider a well-designed GaAs MESFET with 6-dB gain (G_{ma}) at 18 GHz with a 6-dB/octave gain rolloff. The circuit designer must know the maximum bandwidth possible from this transistor using (3.21).

$$\tau_{HP} = \omega_2 R_{in} C_{in} = 0.565$$

Table 3.3   Input Gain Reduction

BW (GHz)	Gain Reduction at f_2 (3-dB/octave slope in M_1)	Gain Reduction at f_2 (6-dB/octave slope in M_1)
9 to 18	0	0
6 to 18	0	0
1.8 to 18	0.3 dB	0
1 to 2	3.3	1.8
1 to 3	3.0	1.6
1 to 10	1.5	0

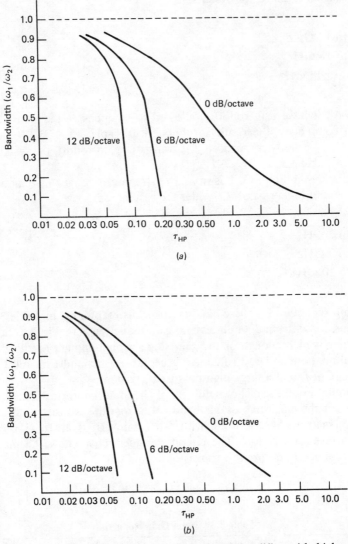

Figure 3.12 Maximum bandwidth for broadband amplifier with high-pass time constant. (*a*) 0-dB loss in G_M at f_2 ($K=1.0$). (*b*) 1-dB loss in G_M at f_2 ($K=0.794$). (*c*) 3-dB loss in G_M at f_2 ($K=0.50$).

114

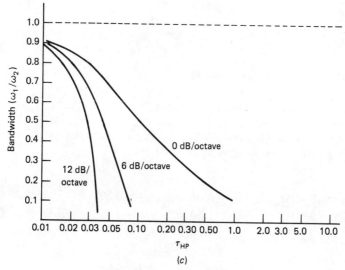

Figure 3.12 (*Continued*)

For a 6-dB/octave rolloff in M_1, Fig. 3.12 gives

$$\frac{\omega_1}{\omega_2} < 0.10$$

so the maximum achievable bandwidth is greater than 1.8 to 18 GHz with a possible gain of 6 dB. The ideal input reflection coefficient is a function of frequency with values of

$$\rho(18) \simeq 0$$

$$\rho(1.8) \simeq 0.99$$

The realization of a broadband input match is a separate problem that might be solved with broadband couplers or power dividers.

The transistor figure of merit is often called the gain-bandwidth product. Since the power gain from (3.12) is

$$G_A = G_0 \left(\frac{f_2}{f} \right)^2 \tag{3.23}$$

the voltage gain is

$$A_v = \sqrt{G_A} = \sqrt{G_0}\, \frac{f_2}{f} \tag{3.24}$$

and the gain bandwidth product is

$$\text{GBW} = \sqrt{G_A}\, f = \sqrt{G_0}\, f_2 \qquad (3.25)$$

For this example the parameter is

$$\text{GBW} = \sqrt{4} \times 18 = 36 \text{ GHz}$$

If there is no output bandwidth limitation, this transistor can be used in the following broadband single-stage amplifiers:

1.8 to 18 GHz	$G_A = 6$ dB	$\tau_{HP} = 0.565$	GBW $= 36$ GHz
1.35 to 9 GHz	$G_A = 12$ dB	$\tau_{HP} = 0.282$	GBW $= 36$ GHz
2.25 to 4.5 GHz	$G_A = 18$ dB	$\tau_{HP} = 0.141$	GBW $= 36$ GHz

For all three of these amplifiers the GBW of the transistor is 36 GHz.

Next, the output gain-bandwidth limitation must be considered for the design in Fig. 3.11. Using the formula in Fig. 3.10a and the following for a lossless reciprocal M_2 network

$$|\Gamma_2|^2 = 1 - |S_{21}|_{\text{out}}^2 = 1 - G_{M2}(f) \qquad (3.26)$$

For a lossless network we have

$$|\Gamma_2| = |\rho_{\text{out}}| \qquad (3.27)$$

$$\int_0^\infty \ln\left|\frac{1}{\rho_{\text{out}}}\right| d\omega \le \frac{\pi}{R_{\text{out}}C_{\text{out}}} \qquad (3.28)$$

Normalizing frequency to $\omega_2 = 1$ gives

$$\Omega = \frac{\omega}{\omega_2} \qquad (3.29)$$

$$\int_0^\infty \ln\left|\frac{1}{\rho_{\text{out}}}\right| d\Omega \le \frac{\pi}{\tau_{LP}} \qquad (3.30)$$

$$\tau_{LP} = \omega_2 R_{\text{out}} C_{\text{out}} \qquad (3.31)$$

$$\int_{\omega_1}^1 \ln\left|\frac{1}{\rho_{\text{out}}}\right| d\Omega \le \frac{\pi}{\tau_{LP}} \qquad (3.32)$$

Thus, ρ_{out} cannot be zero over a finite bandwidth; however, the wider the bandwidth, the higher ρ_{out} can become. Notice that for the time constant τ_{LP} a low value is now required; this is the opposite condition than for the input high-pass circuit. If zero shunt capacitance could be achieved, there is no gain-bandwidth restriction. From (3.14), (3.26), and (3.27) we get

$$\int_{\omega_1}^{1} \ln \left| \frac{1}{1 - K(f/f_2)^k} \right| d\Omega \le \frac{2\pi}{\tau_{LP}} \tag{3.33}$$

The integration of (3.33) is plotted in Fig. 3.13 to indicate the gain-bandwidth restrictions of the output. As an example to illustrate the use of Fig. 3.13, consider a FET chip with the following model:

$$R_{out} = 500$$

$$C_{out} = 0.10 \text{ pF}$$

For the same wideband designs considered before, the calculated output gain bandwidth limits are tabulated in Table 3.4.

Now lower-frequency amplifiers have no gain-bandwidth reduction, while the high-frequency broadband amplifiers have a gain bandwidth

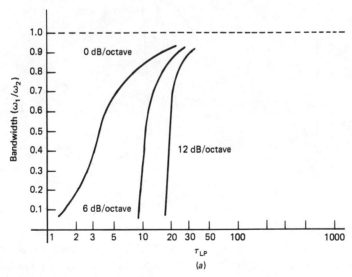

Figure 3.13 Maximum bandwidth for broadband amplifier with low-pass time constant. (*a*) 0-dB loss in G_M at f_2 ($K = 1.0$). (*b*) 1-dB loss in G_M at f_2 ($K = 0.794$). (*c*) 3-dB loss in G_M at f_2 ($K = 0.50$).

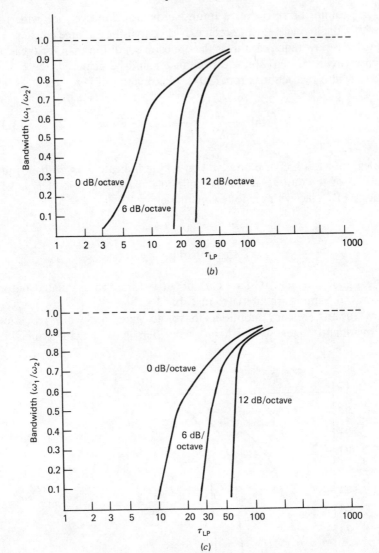

Figure 3.13 (*Continued*)

limitation unless the gain of M_2 is sloped about 3 dB. From the model of the transistor and the results of Figs. 3.12 and 3.13, the designer can distribute the gain slopes of M_1 and M_2 for maximum bandwidth. For chip GaAs FET amplifiers, the input circuit is usually used for gain sloping.

Another example of the use of Fig. 3.13 is the maximum expected bandwidth from the model with a 0-dB gain slope. Consider a transistor

Table 3.4 Output Gain Reduction

BW (GHz)	Gain Reduction at f_2 (0-dB/octave slope in M_2)	Gain Reduction at f_2 (3-dB/octave in M_2)	Gain Reduction at f_2 (6-dB/octave in M_2)
9 to 18	0.50	0	0
6 to 18	1.0	0	0
1.8 to 18	2.0	0.1	0
1 to 2	0	0	0
1 to 3	0	0	0
1 to 10	0.10	0	0

with a 6-dB gain at 18 GHz with a 6-dB/octave gain slope and the previous output model. Since the input was used for gain sloping, the output network M_2 must give a 0-dB slope. From (3.31)

$$\tau_{LP} = \omega_2 R_{out} C_{out} = 5.65$$

For a 0-dB slope in M_2, Fig. 3.13a gives

$$\frac{\omega_1}{\omega_2} = 0.70$$

So the maximum achievable bandwidth is 12.6 to 18 GHz with a gain of 6 dB. If a 1-dB reduction in gain is acceptable, Fig. 3.13b gives

$$\frac{\omega_1}{\omega_2} = 0.20$$

So the maximum achievable bandwidth is 3.6 to 18 GHz with a gain of 5 dB. Alternatively, if a 6-dB/octave slope is allowed, Fig. 3.13a gives

$$\frac{\omega_1}{\omega_2} = 0.10$$

So the maximum achievable bandwidth is 1.8 to 18 GHz. For this transistor, some gain slope should be assigned for both input and output (M_1 and M_2) if the design band is 2 to 18 GHz.

An acceptable wideband GaAs MESFET chip must have three conditions:

1 High gain at f_2 (G_{ma} or G_{ms}).
2 High value of $\tau_{HP} = \omega_2 R_{in} C_{in}$.
3 Low value of $\tau_{LP} = \omega_2 R_{out} C_{out}$.

The bandwidth limitations are given by Fig. 3.12 and 3.13, which are Fano's limit for broadband matching with a gain slope included. The distribution of gain slopes and possible overall gain reduction is determined from these curves. The gain-bandwidth figure of merit is given by (3.25) for a transistor with a 6-dB/octave power gain rolloff.

3.6   Feedback Design

The addition of feedback to the transistor amplifier should be considered for several applications:

1 Improvement of input match S_{11} over frequency.
2 Improvement of output match S_{22} over frequency.
3 Flatten transducer gain S_{21} over frequency.
4 Improve stability factor k by reducing S_{12} over frequency.
5 Cause S_{11} or S_{22} to exceed unity by increasing S_{12} for an oscillator application.

Feedback is a useful technique for modifying the transistor S-parameters before designing the amplifier (or oscillator) passive networks. The first four applications apply to amplifiers, whereas the fifth is for oscillators.

The most common types of feedback are shown in Fig. 3.14. The series feedback is often used to improve S_{11} at the expense of reduced stability. The parallel feedback can be used to flatten gain over frequency. For series feedback or current series feedback, the z-parameters of the two two-ports add. For parallel feedback or voltage shunt feedback, the y-parameters of the two two-ports add. Since the result of feedback must be understood in S-parameters, it is often helpful to map the result of feedback. For example, the case of series feedback on S_{11} is plotted for a typical GaAs MESFET chip at 8 GHz in Fig. 3.15. The overall effect on the other S-parameters and the gain must also be considered before using feedback to improve S_{11}.

The addition of resistive feedback to the transistor is usually undesired because of reduced gain, higher noise figure, and reduced output power capability. However, for high-gain bipolar transistors, a commonly used design is compound feedback, both series and shunt resistive feedback, as shown in Fig. 3.16. This type of amplifier includes both series and parallel feedback with reactive elements to peak the high-frequency gain.

Normally, R_f and R_s are chosen to satisfy

$$Z_0^2 = R_f R_s \qquad\qquad (3.34)$$

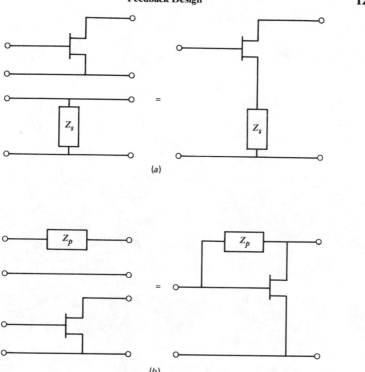

Figure 3.14 Transistor feedback. (*a*) Series feedback; (*b*) Parallel feedback.

which gives

$$Z_{\text{in}} \simeq Z_{\text{out}} \simeq 50 \text{ ohms} \tag{3.35}$$

Notice that a series inductor with R_f removes this resistor from the circuit at high frequencies; also, a shunt capacitor with R_s removes this resistor at high frequencies. Usually, R_1 is added for dc biasing and is a large value, typically greater than 300 ohms. By proper adjustment of R_f and R_s in combination with reactive elements, a flat S_{21} can be achieved over a wide bandwidth with acceptable S_{11} and S_{22}. The same technique does not work well for the GaAs MESFET, because of the lower value of S_{21} at low frequencies.

Another feedback technique for realizing the maximum possible gain from a transistor is unilaterization of the transistor. By adding either shunt or series feedback, the value of S_{12} can be resonated to zero at a single frequency. However, over a broad frequency range, the value of S_{12} will not

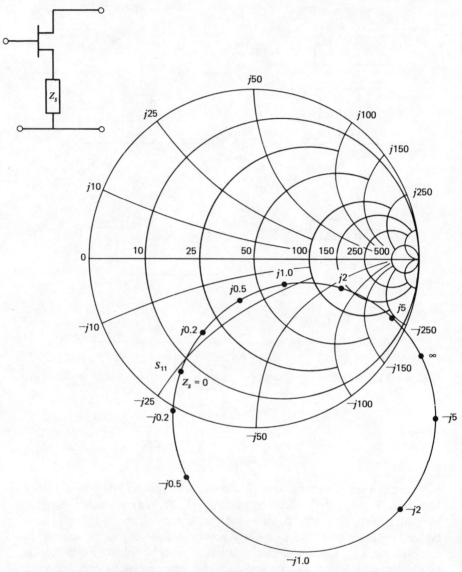

Figure 3.15 Series feedback for GaAs MESFET input reflection coefficient. Coordinates in ohms. HFET-1001 chip data, $V_{DS} = 4.0$ V, $I_{DS} = I_{DSS}$.

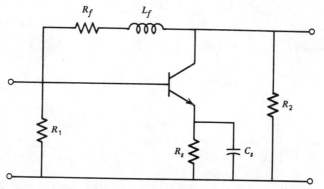

Figure 3.16 Series and parallel feedback for wideband bipolar transistor amplifiers.

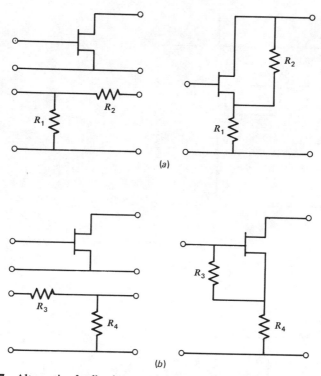

(a)

(b)

Figure 3.17 Alternative feedback networks. (a) Series-parallel feedback or voltage-series feedback; (b) parallel-series feedback or current-shunt feedback.

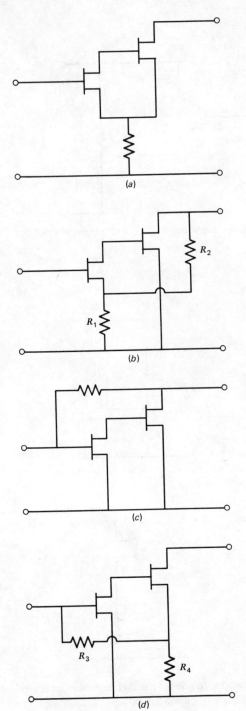

3.18 Two-stage feedback. (*a*) Series; (*b*) Series-parallel; (*c*) Parallel; (*d*) Parallel-series.

124

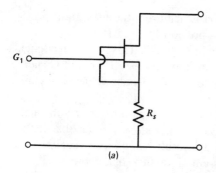

(a)

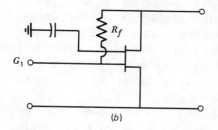

(b)

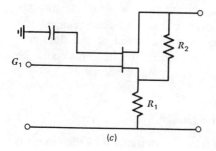

(c)

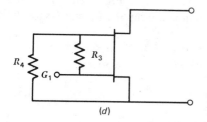

(d)

3.19 Dual-gate feedback networks. (*a*) Series; (*b*) Parallel; (*c*) Series-parallel; (*d*) Parallel-series.

125

equal zero. When the unilateral condition is achieved, the maximum possible gain reaches the unilateral gain given previously by (1.165).

Two other types of feedback are given in Fig. 3.17. This type of feedback can be used to adjust the input and output impedances differently. Usually series feedback increases impedance while parallel feedback decreases impedance.

For the case of series-parallel feedback, the h-parameters of the networks add directly. For the opposite case of parallel-series feedback, the g-parameters add directly. Since the final result should be visualized in S-parameters, a computer is nearly essential to investigate the feedback networks in Figs. 3.14 and 3.17.

An additional form of feedback is two-stage feedback, as shown in Fig. 3.18. This feedback may be used for either amplifier or oscillator design. Since transistors can be connected as common-source (CS), common-gate (CG), and common-drain (CD), there are a large number of combinations which can be investigated for different applications. Usually because of economics and because there has been no significant improvement in performance, the single-stage designs are preferred.

The application of feedback to the dual-gate MESFET is shown in Fig. 3.19. Since this transistor is a two-stage transistor, the feedback networks in Fig. 3.18 can be applied to the two-stage combinations of CS–CG. For broadband amplifiers, the series-parallel combination is an interesting approach to reducing both S_{11} and S_{22} over a wide frequency range. The parallel feedback is not recommended, since the virtue of the dual-gate MESFET is a very low S_{12}. Since the termination at gate 2 may also be arbitrarily chosen, a large number of two-port circuits are possible.

3.7 Two-Stage Design

If the gain per stage is low or the physical size requirement of the circuit is very small, a two-stage amplifier design should be considered. The general configuration for common-source stages is given in Fig. 3.20, where the passive network design requirements now include input M_1, interstage M_2, and output M_3.

For a high-gain stage, the lossless network M_2 is designed to give a conjugate power match between stages, which can be expressed by:

$$(S_{11})_2 = (S_{22})_1^* \tag{3.36}$$

$$(S_{22})_1 = (S_{11})_2^* \tag{3.37}$$

where the assumption that $S_{12} = 0$ is used for simplification. Thus the design

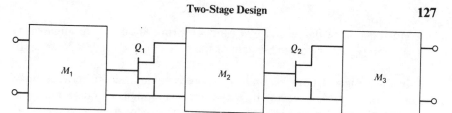

Figure 3.20   Two-stage amplifier design.

of a two-stage amplifier is basically the same as for the single-stage amplifier in Fig. 1.1 with the additional step of designing an interstage network M_2 that satisfies either (3.36) or (3.37). If M_2 is lossless, the conditions expressed by (3.36) and (3.37) are identical.

The two-port stability of the two-stage amplifier must be considered separately. Since two active devices are present, the stability factor of the two-stage amplifier does not guarantee overall stability. If the two-port k is greater than unity, there is no passive network connected to either M_1 or M_3 that can make the amplifier oscillate. However, the interstage network M_2 could cause the transistor to oscillate if either individual transistor stability factor is less than unity. At low frequencies, where the transistor stability factor is less than unity, the termination provided by the interstage network must be checked for stability of both Q_1 and Q_2. Since biasing stubs are required for Q_1 and Q_2, these shorted stubs usually guarantee stability in the interstage network.

For two-stage low noise design, the requirement on M_2 is no longer the conjugate power match given by (3.36) or (3.37). The requirement for M_1 is:

$$(\Gamma_G)_1 = \Gamma_{on1} \tag{3.38}$$

and for M_2 is:

$$(\Gamma_G)_2 = \Gamma_{on2} \tag{3.39}$$

The design of M_3 is for a conjugate power match at the output. These conditions should be achieved from lossless matching networks, since resistors will add noise and reduce gain. Normally, a premium low-noise transistor is used in the first stage and a slightly higher noise figure in the second stage. From (3.2), the total noise figure of the amplifier is

$$F = F_1 + \frac{F_2 - 1}{G_{av1}} \tag{3.40}$$

where F_1 is the first-stage noise figure and is a function of M_1, F_2 is the second-stage noise figure and is a function primarily of M_2 terminated by $(S_{22})_1'$, and G_{av1} is the available gain of the first stage, which is a function of M_1.

In the design of both M_1 and M_2, there is some tradeoff between noise figure and gain, but normally for low-noise design, a noise match is required. The other important considerations are overall stability at all frequencies and flat gain over the operating bandwidth.

In a two-stage high-power design, the requirement for M_2 is

$$(\Gamma_L)_1 = \Gamma_{op1} \tag{3.41}$$

and for M_3 is

$$(\Gamma_L)_2 = \Gamma_{op2} \tag{3.42}$$

The design of M_1 is for a conjugate power match at the input. These conditions should be achieved from lossless matching networks, since resistors will absorb part of the output power that is intended for the load.

The output transistor is now the higher-cost transistor, which must deliver the required output power. After calculating the large-signal gain of the output transistor, the input transistor may be selected. The overall transducer gain is the total large-signal gain of each stage (in dB):

$$G_T = G_{LS1} + G_{LS2} \tag{3.43}$$

where G_{LS1} is primarily a function of M_2 terminated by $(S_{11})_2'$ and G_{LS2} is a function of M_3 and M_2 terminated by $(S_{22})_1'$.

The interstage design requirements for high gain, low noise, and high power are summarized in Fig. 3.21. It is helpful to visualize the interstage design so that an impedance matching network moves from the load to the generator. This is consistent with the impedance matching techniques discussed in Section 2.5. Thus the low-noise interstage design consists of moving an imaginary load of Γ_{on2}^* to the point $(S_{22})_1^*$ on the Smith Chart.

The design of broadband two-stage amplifiers requires an overall understanding of the gain-bandwidth limitations discussed in Section 3.5. In addition to the input and output gain-bandwidth limitations, the interstage has additional limitations, as shown in Fig. 3.22 for a GaAs MESFET two-stage amplifier. Since there are two parasitic capacitors that must be absorbed in the matching structure, the design is not straightforward. The desired performance from M_2 is a conjugate power match at f_2 and a gain slope that may be between 0 and 12 dB/octave depending on the slopes required at M_1 and M_3. The total gain slope of M_1, M_2, and M_3 is typically 12 dB/octave, since two transistor gain slopes must be compensated for. A

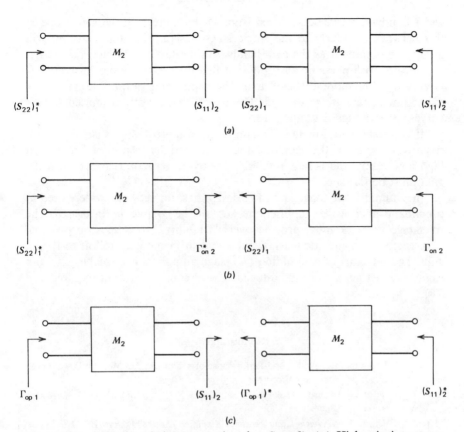

Figure 3.21 Interstage designs (assuming that $S_{12} = 0$). (a) High-gain interstage; (b) Low-noise interstage; (c) High-power interstage.

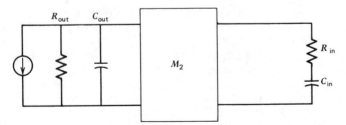

Figure 3.22 Interstage matching network for GaAs MESFET amplifier.

wider bandwidth can be achieved from M_2 by increasing the gain slope in this network. Similarly to the single stage design, one time constant will usually be dominant as the gain-bandwidth limitation. The network design then begins with parasitic absorption at that port. Usually the design of M_2 is considerably more difficult than the input and output design if the maximum bandwidth is desired. The overall bandwidth is limited by the curves given in Figs. 3.12 and 3.13.

If a broadband low-noise amplifier is required, M_2 is set by noise requirements. Now the gain must be flattened by M_3. If a broadband high-power amplifier is required, M_2 is set by power requirements. Now the gain must be flattened by M_1.

In summary, two-stage amplifier design has not yet been widely used in broadband microwave amplifiers because of the complexity in design. The interstage network must provide overall stability, parasitic absorption of two reactive elements, dc biasing stubs, and sufficient gain rolloff to flatten gain. The advantages of building two-stage amplifiers are smaller size and slightly higher gain, since the interstage 3-dB couplers are not required.

Bibliography

3.1 G. R. Basawapatna, "Design and Performance of a 2 to 18 GHz Medium Power GaAs FET Amplifier," *European Microwave Conference*, 1978.

3.2 L. Besser and S. Swenson, "Take the Hassle Out of FET Amplifier Design," *Microwave Systems News*, September 1977, pp. 97–105.

3.3 W. K. Chen, *Active Network and Feedback Amplifier Theory*, McGraw-Hill, New York, 1980.

3.4 J. V. DiLorenzo and W. R. Wisseman, "GaAs Power MESFET's: Design, Fabrication, and Performance," *IEEE Trans. on MTT*, Vol. MTT-23, May 1979, pp. 367–378.

3.5 R. M. Fano, "Theoretical Limitations on the Broadband Matching of Arbitrary Impedances," *Journal of the Franklin Institute*, Vol. 249, January 1960, pp. 57–83, and February 1960, pp. 139–155.

3.6 M. G. Fornaciari, "Low Noise Amplifier Design for Satellite Receivers," *1979 WESCON Convention Digest*, August 1979, Session 25.

3.7 W. H. Ku and W. C. Petersen, "Optimum Gain-Bandwidth Limitations of Transistor Amplifiers as Reactively Constrained Active Two-Port Networks," *IEEE Transactions on Circuits and Systems*, Vol. CAS-22, June 1975, pp. 523–533.

3.8 J. Lange, "Interdigitated Stripline Quadrature Hybrid," *IEEE Transactions on MTT*, Vol. MTT-17, December 1969, pp. 1150–1151.

3.9 C. A. Liechti and R. L. Tillman, "Design and Performance of Microwave Amplifiers with GaAs Schottky-Gate Field-Effect Transistors," *IEEE Transactions on MTT*, Vol. MTT-22, May 1974, pp. 510–517.

3.10 D. J. Mellor, *Computer-Aided Synthesis of Matching Networks for Microwave Amplifiers*, Stanford Electronics Laboratories SEL-75-005 (Ph.D. Thesis), March 1975.

3.11 D. J. Mellor and J. G. Linvill, "Synthesis of Interstage Networks of Prescribed Gain Versus Frequency Slopes," *IEEE Transactions on Microwave Theory and Techniques*, Vol. MTT-23, December 1975, pp. 1013–1020.

3.12 K. B. Niclas, W. T. Wilser, R. B. Gold, and W. R. Hitchens, "The Matched Feedback Amplifier: Ultrawide-Band Microwave Amplification with GaAs MESFET's," *IEEE Transactions on MTT*, Vol. MTT-28, April 1980, pp. 285–294.

3.13 K. B. Niclas, "Compact Multistage Single-Ended Amplifiers for S, C, and X Band Operation," *1981 International Microwave Symposium*, June 1981, pp. 132–134.

3.14 K. B. Niclas and W. T. Wilser, "A 2–12 GHz Feedback Amplifier on GaAs," *1981 International Microwave Symposium*, June 1981, pp. 356–358.

3.15 R. S. Tucker, "Gain-Bandwidth Limitations of Microwave Transistor Amplifiers," *IEEE Transactions on Microwave Theory and Techniques*, Vol. MTT-21, May 1973, pp. 322–327.

3.16 E. Ulrich, "Use Negative Feedback to Slash Wideband VSWR," *Microwaves*, October 1978, pp. 66–70.

3.17 G. D. Vendelin, "Feedback Effects on GaAs MESFET Noise Performance," *1975 International Microwave Symposium*, May 1975, pp. 324–326.

3.18 G. D. Vendelin, "Feedback Effects in the GaAs MESFET Model," *IEEE Transactions on MTT*, June 1976, pp. 383–385.

3.19 G. D. Vendelin, "2–4 GHz FETs Compete with Bipolar for Low Noise Designs," *Microwave Systems News*, January 1977, pp. 71–75.

3.20 G. D. Vendelin, "Power GaAs FET Amplifier Design with Large Signal Tuning Parameters," *11th Asilomar Conference on Circuits, Systems, and Computers*, November 1977, pp. 139–141.

3.21 G. D. Vendelin, "Computer-Aided Design of Broadband GaAs MESFET Microwave Integrated Circuits," *Military Electronics Defense Expo '79 Conference Proceedings*, September 1979, pp. 149–157.

CHAPTER FOUR

OSCILLATOR DESIGNS

4.0 Introduction

Oscillator design is very similar to amplifier design. The same transistors, the same dc bias levels, and the same set of S-parameters can be used for the oscillator design. The load does not know whether it is connected to an oscillator or an amplifier (see Fig. 1.1).

For the amplifier design, M_1 and M_2 can be designed with a normal Smith Chart, since S'_{11} and S'_{22} are normally less than unity. For oscillators, S'_{11} and S'_{22} are both greater than unity for oscillation. Thus a compressed Smith Chart that includes reflection coefficients greater than unity is a useful tool for oscillator design.

The conditions for oscillation can be expressed as

$$k < 1 \tag{4.1}$$

$$\Gamma_G S'_{11} = 1 \tag{4.2}$$

$$\Gamma_L S'_{22} = 1 \tag{4.3}$$

The stability factor should be less than unity for any possibility of oscillation. If this condition is not satisfied, either the common terminal should be changed or positive feedback should be added. Next the passive terminations Γ_G and Γ_L must be added to resonate the input and output ports at the frequency of oscillation. This is satisfied by either (4.2) or (4.3). It will be shown in Section 4.3 that if (4.2) is satisfied, (4.3) must be satisfied and vice versa. In other words, if the oscillator is oscillating at one port, it must be simultaneously oscillating at the other port. Normally a major fraction of the power is delivered only to one port, since only one load is connected. Since $|\Gamma_G|$ and $|\Gamma_L|$ are less than unity, (4.2) and (4.3) imply that $|S'_{11}| > 1$ and $|S'_{22}| > 1$.

The conditions for oscillation can be seen from Fig. 4.1, where an input generator has been connected to a two-port. Using (1.78) for the representation of the generator, which is repeated here:

$$a_1 = b_G + \Gamma_1 \Gamma_G a_1 \tag{4.4}$$

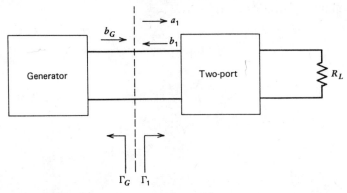

Figure 4.1 Two-port connected to a generator.

and defining

$$\Gamma_1 = S'_{11} \qquad (4.5)$$

$$S'_{11} = \frac{b_1}{a_1} \qquad (4.6)$$

gives

$$b_G = a_1(1 - \Gamma_1 \Gamma_G)$$

$$= \frac{b_1}{S'_{11}}(1 - S'_{11}\Gamma_G) \qquad (4.7)$$

$$\frac{b_1}{b_G} = \frac{S'_{11}}{1 - S'_{11}\Gamma_G} \qquad (4.8)$$

Thus the wave reflected from the two-port is dependent on b_G, S'_{11}, and Γ_G. If (4.2) is satisfied, b_G must be zero, which implies that the two-port is oscillating. Since normally $|\Gamma_G|$ is less than or equal to unity, this requires that $|S'_{11}|$ be greater than or equal to unity.

The oscillator designer must simply guarantee a stability factor less than unity and resonate the input port by satisfying (4.2), which that implies (4.3) has also been satisfied. Another way of expressing the resonance condition of (4.2) is the following:

$$R_{in} + R_G = 0 \qquad (4.9)$$

$$X_{in} + X_G = 0 \qquad (4.10)$$

This follows from substituting

$$S'_{11} = \frac{R_{in} + jX_{in} - Z_0}{R_{in} + jX_{in} + Z_0} \tag{4.11}$$

$$\Gamma_G = \frac{R_G + jX_G - Z_0}{R_G + jX_G + Z_0}$$

$$= \frac{-R_{in} - Z_0 - jX_{in}}{-R_{in} + Z_0 - jX_{in}} \tag{4.12}$$

into (4.2), giving

$$\Gamma_G S'_{11} = \frac{-R_{in} - Z_0 - jX_{in}}{-R_{in} + Z_0 - jX_{in}} \cdot \frac{R_{in} + jX_{in} - Z_0}{R_{in} + jZ_0 + jX_{in}} = 1$$

which proves the equivalence of (4.2) to (4.9) and (4.10).

Table 4.1   Typical Oscillator Specifications

Parameter	High-Q or Cavity-Tuned (e.g., YIG)	Low-Q or Varactor-Tuned VCO
Frequency	2 to 4 GHz	2 to 4 GHz
Power	+10 dBm	+10 dBm
Power variation versus f	±2 dB	±2 dB
Temperature stability versus f	±10 ppm/°C	±500 ppm/°C
Power versus temperature (−30 to 60°C)	±2 dB	±2 dB
Modulation sensitivity	10 to 20 MHz/mA	50 to 200 MHz/V
FM noise	−110 dBc/Hz @100 KHz	−100 dBc/Hz @100 KHz
AM noise	−140 dBc/Hz @100 KHz	−140 dBc/Hz @100 KHz
FM noise floor	−150 dBc/Hz @100 MHz	−150 dBc/Hz @100 MHz
All harmonics	−20 dBc	−20 dBc
Short-term post tuning drift	±2 MHz 1 μsec	±2 MHz 1 to 100 μsec
Long-term post tuning drift	±2 MHz 5 to 30 sec	±2 MHz 5 to 30 sec
Pulling of f all phases of 12-dB return loss	±1 MHz	±20 MHz
Pushing of f with change of bias voltage	5 MHz/V	5 MHz/V

Before proceeding with the oscillator design procedures, some typical oscillator specifications are given in Table 4.1 for the major types of oscillators. The high-Q or cavity-type oscillators usually have better spectral purity (see Section 4.4) compared to the low-Q VCO's (voltage-controlled oscillators), which have faster tuning speeds. The YIG is a material that provides a high-Q resonance in a magnetic field; YIG means yttrium iron garnet, which is $Y_2Fe_2(FeO_4)_3$. The FM noise is usually measured at about 100 kHz from the carrier in units of dBc, which means decibels below the carrier level, in a specified bandwidth of 1 Hz. If the measurement bandwidth is 1 kHz, the specification changes by 10^3 (see Section 4.4).

In selecting a transistor to meet the specifications, the amplifier transistors with the same frequency and power performance are usually suitable. Lower close-in noise can be achieved from Si bipolar transistors compared to GaAs MESFETs because of the $1/f$ noise difference described in Section 4.4.

4.1   The Compressed Smith Chart

The normal Smith Chart is a plot of reflection coefficient of $|\Gamma| \leq 1$. The compressed Smith Chart includes $|\Gamma| > 1$, and the chart is given in Fig. 4.2 for $|\Gamma| \leq 3.16$ (10 dB of return gain). This chart is useful for plotting the variation of S'_{11} and S'_{22} for oscillator design. The impedance and admittance properties of the Smith Chart are retained for the compressed chart. For example, a Γ_{in} of $1.2 \angle 30°$ gives the following values of Z and Y normalized to $Z_0 = 50$ ohms.

$$\frac{Z_{in}}{Z_0} = -0.10 + j0.25$$

$$\frac{Z_{in}^*}{Z_0} = -0.10 - j0.25$$

$$\frac{Y_{in}}{Y_0} = -1.0 - j3.0$$

$$\frac{Y_{in}^*}{Y_0} = -1.0 + j3.0$$

These values are plotted in Fig. 4.2 for illustration.

A frequency resonance condition simply requires the circuit imaginary term be zero. If the impedance resonance is on the left-hand real axis, this is a series resonance; that is, at frequencies above resonance the impedance is inductive and below resonance the impedance is capacitive. If the impedance resonance is on the right-hand real axis, the resonance is a parallel

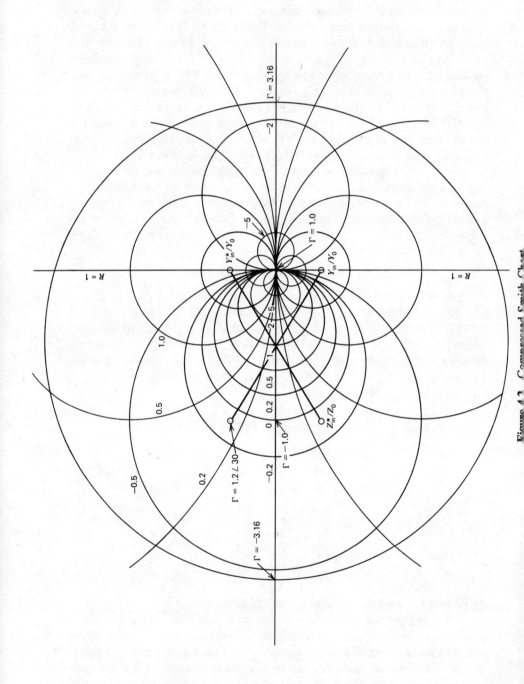

Figure 4.3 Compressed Smith Chart

136

resonance; that is, at frequencies above resonance the impedance is capacitive and below resonance the impedance is inductive.

An oscillator resonance condition implies that both the circuit imaginary term and the circuit real term are zero, as given by (4.9) and (4.10). Impedances and admittances can be transformed on the compressed Smith Chart by the same methods discussed in Section 2.5; however, when $|\Gamma|$ is greater than unity, the goal of impedance transformation is usually to achieve either a series or a parallel resonance condition.

4.2   Series or Parallel Resonance

Oscillators can be classified into two types, series-resonant or parallel-resonant, as shown in Fig. 4.3. The equivalent circuit of the active device is chosen from the frequency response of the output port, that is the frequency response of Γ_G. For the series-resonant condition, the negative resistance of

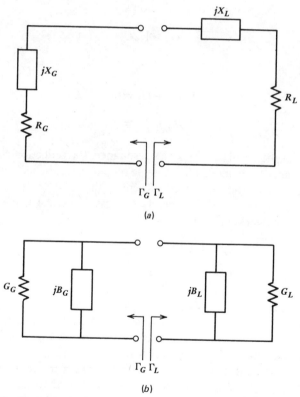

Figure 4.3   Oscillator-equivalent circuits. (*a*) Series-resonant; (*b*) Parallel-resonant.

the active device must exceed the load resistance R_L at start-up of oscillation by about 20%. As the oscillation builds up to a steady-state value, the resonance condition will be reached as a result of limiting effects, which cause a reduction of R_G under large-signal drive.

For start-up of oscillation

$$R_G > 1.2 R_L \tag{4.13}$$

for resonance

$$R_G + R_L = 0 \tag{4.14}$$

$$X_G + X_L = 0 \tag{4.15}$$

For the parallel resonant condition, the negative conductance of the active device must exceed the load conductance G_L at start-up of oscillation by about 20%. The parallel resonant oscillator is simply the dual of the series resonant case. For start up of oscillation

$$G_G > 1.2 G_L \tag{4.16}$$

For resonance

$$G_G + G_L = 0 \tag{4.17}$$

$$B_G + B_L = 0 \tag{4.18}$$

To design the oscillator for series resonance, the reflection coefficient of the active transistor is moved to an angle of 180° (i.e., the left-hand real axis of the compressed Smith Chart). Keeping in mind (4.2) for the input resonating port, we see that a nearly lossless reactance will resonate the transistor. For the example in Fig. 4.2,

$$\Gamma_G = 1.2 \angle 30°$$

$$\Gamma_L = 0.83 \angle -30° \simeq 1.0 \angle -30°$$

If Γ_L is lossless, the large-signal drive of the transistor will reduce Γ_G to about $1.0 \angle 30°$. For parallel resonance oscillator design, the reflection coefficient of the active transistor is moved to an angle of 0° (i.e., the right-hand real axis of the compressed Smith Chart). Alternatively, the reflection coefficient associated with impedance can be inverted to an admittance point, and the admittance can be moved to an angle of 180° (i.e., the left-hand real axis of the compressed Smith Chart).

The passive resonator need not be the same type as the active device. For example, a series-resonant device may be brought to resonance by a parallel-resonant circuit. An example of this is a bipolar transistor YIG-tuned oscillator.

4.3   Two-Port Oscillator Design

A common method for designing oscillators is to resonate the input port with a passive high-Q circuit at the desired frequency of resonance. It will be shown that if this is achieved with a load connected on the output port, the transistor is oscillating at both ports and is thus delivering power to the load port. The oscillator may be considered a two-port structure as shown in Fig. 1.1, where M_3 is the lossless resonating port and M_4 provides lossless matching such that all of the external rf power is delivered to the load. The resonating network could be a YIG, a varactor, a dielectric resonator, a lumped element tuning network, or a microstripline tuning network. Normally only parasitic resistance is present at the resonating port, since a high-Q resonance is desirable for minimizing oscillator noise. It is possible to have loads at both the input and the output ports if such an application occurs, since the oscillator is oscillating at both ports simultaneously.

The simultaneous oscillation condition is proved as follows. Assume that the oscillation condition is satisfied at port 1:

$$\frac{1}{S'_{11}} = \Gamma_G \tag{4.19}$$

From (1.36)

$$S'_{11} = S_{11} + \frac{S_{12}S_{21}\Gamma_L}{1 - S_{22}\Gamma_L} = \frac{S_{11} - D\Gamma_L}{1 - S_{22}\Gamma_L} \tag{4.20}$$

$$\frac{1}{S'_{11}} = \frac{1 - S_{22}\Gamma_L}{S_{11} - D\Gamma_L} = \Gamma_G \tag{4.21}$$

From expanding (4.21) we get

$$\Gamma_G S_{11} - D\Gamma_L\Gamma_G = 1 - S_{22}\Gamma_L$$

$$\Gamma_L(S_{22} - D\Gamma_G) = 1 - S_{11}\Gamma_G$$

$$\Gamma_L = \frac{1 - S_{11}\Gamma_G}{S_{22} - D\Gamma_G} \tag{4.22}$$

From (1.38)

$$S'_{22} = S_{22} + \frac{S_{12}S_{21}\Gamma_G}{1 - S_{11}\Gamma_G} = \frac{S_{22} - D\Gamma_G}{1 - S_{11}\Gamma_G} \tag{4.23}$$

$$\frac{1}{S'_{22}} = \frac{1 - S_{11}\Gamma_G}{S_{22} - D\Gamma_G} \tag{4.24}$$

Comparing (4.22) and (4.24) gives

$$\frac{1}{S'_{22}} = \Gamma_L \tag{4.25}$$

which means that the oscillation condition is also satisfied at port 2; this completes the proof. Thus if either port is oscillating, the other port must be oscillating as well. A load may appear at either or both ports, but normally the load is in Γ_L, the output termination.

An oscillator may be viewed as an amplifier that delivers power to the output load and to the input impedance of the amplifier. There are four degrees of freedom in the oscillator design, the reactance presented at each of the three terminals and the load resistance required for oscillation. The load may be placed in series or parallel with any of the reactive elements. The oscillator can be viewed as a common source (or common emitter) with either series resonance at all terminals or parallel resonance at all ports, as shown in Fig. 4.4.

The circuits in Fig. 4.4 reduce to Fig. 4.5 when the load is added. There are six forms the oscillator may take with simple resonating structures, as shown in Fig. 4.5. The first three oscillator structures may be viewed as resonant circuits with series feedback. The second three structures are parallel-resonant with parallel feedback. The ground has not been specified, although usually the load is grounded. All of these oscillators can be analyzed with two-port network theory. The best choice will depend on practical realizability and dc biasing considerations.

It is helpful to use the common-source amplifier to compute the oscillator output power. For oscillators, the objective is to maximize $(P_{out} - P_{in})$ of the amplifier, which is the useful power to the load. An empirical expression for the common-source amplifier output power found by Johnson is

$$P_{out} = P_{sat}\left(1 - \exp\frac{-GP_{in}}{P_{sat}}\right) \tag{4.26}$$

where P_{sat} is the saturated output power of the amplifier and G is the tuned

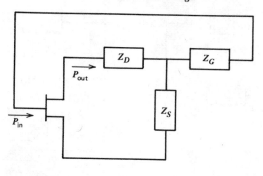

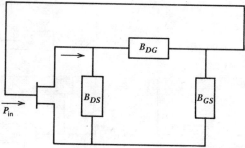

Figure 4.4 Common-source amplifier connected to T or Pi coupling network to form oscillator.

small-signal common-source transducer gain of the amplifier, which is identical to $|S_{21}|^2$. Since the objective is to maximize $P_{out} - P_{in}$,

$$d(P_{out} - P_{in}) = 0 \tag{4.27}$$

$$\frac{\partial P_{out}}{\partial P_{in}} = 1 \tag{4.28}$$

$$\frac{\partial P_{out}}{\partial P_{in}} = G\exp - \frac{GP_{in}}{P_{sat}} = 1 \tag{4.29}$$

$$\exp\frac{GP_{in}}{P_{sat}} = G \tag{4.30}$$

$$\frac{P_{in}}{P_{sat}} = \frac{\ln G}{G} \tag{4.31}$$

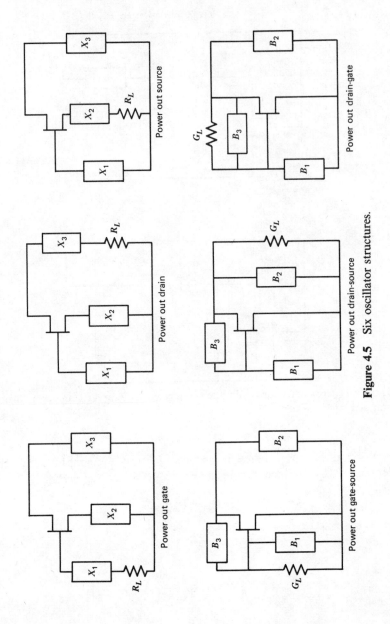

Figure 4.5 Six oscillator structures.

142

At the maximum value of $P_{out} - P_{in}$, the amplifier output is

$$P_{out} = P_{sat}\left(1 - \frac{1}{G}\right) \tag{4.32}$$

and the maximum oscillator output power is

$$P_{osc} = P_{out} - P_{in}$$

$$= P_{sat}\left(1 - \frac{1}{G} - \frac{\ln G}{G}\right) \tag{4.33}$$

Thus the maximum oscillator output power can be predicted from the common-source amplifier saturated output power and the small-signal common source transducer gain G. A plot of (4.33) is given in Fig. 4.6,

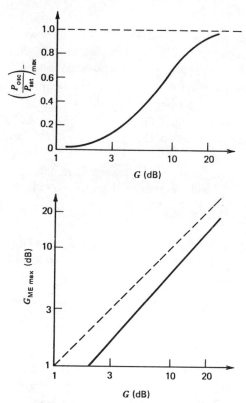

Figure 4.6 Maximum oscillator power and maximum efficient gain versus small-signal transducer power gain.

which shows the importance of high gain for a high oscillator output power. Another gain that is useful for large-signal amplifier or oscillator design is the maximum efficient gain, defined by

$$G_{\mathrm{ME}} = \frac{P_{\mathrm{out}} - P_{\mathrm{in}}}{P_{\mathrm{in}}} \tag{4.34}$$

For maximum oscillator power the maximum efficient gain is, from (4.31) and (4.32),

$$G_{\mathrm{MEmax}} = \frac{G-1}{\ln G} \tag{4.35}$$

This gain is also plotted in Fig. 4.6, showing a considerably smaller value of G_{MEmax} compared to G, the small-signal gain.

Two-port oscillator design may be summarized as follows:

1 Select transistor with sufficient gain and output power capability for the frequency of operation. This may be based on oscillator data sheets, amplifier performance, or S-parameter calculation.

2 Select a topology that gives $k < 1$ at the operating frequency. Add feedback if $k < 1$ has not been achieved.

3 Select an output load matching circuit that gives $|S'_{11}| > 1$ over the desired frequency range. In the simplest case this could be a 50-ohm load.

4 Resonate the input port with a lossless termination so that $\Gamma_G S'_{11} = 1$. The value of S'_{22} will be greater than unity with the input properly resonated.

The transistor will oscillate with any of the six configurations given in Fig. 4.5. In all cases the transistor delivers power to a load and the input of

Table 4.2   HP2001 Bipolar Chip Common-Base
($V_{\mathrm{CE}} = 15$ V, $I_C = 25$ mA)

$L_B = 0$	$L_B = 0.5$ nH	
$S_{11} = 0.94 \angle 174°$	1.04	$\angle 173°$
$S_{21} = 1.90 \angle -28°$	2.00	$\angle -30°$
$S_{12} = 0.013 \angle 98°$	0.043	$\angle 153°$
$S_{22} = 1.01 \angle -17°$	1.05	$\angle -18°$
$k = -0.09$		-0.83

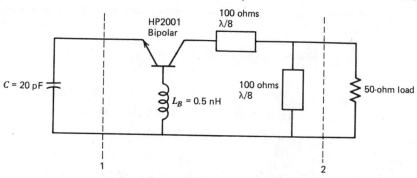

Figure 4.7 Oscillator example at 2 GHz.

the transistor. Practical considerations of realizability and dc biasing will determine the best design.

For both bipolar and FET oscillators, a common topology is common-base or common-gate, since a common-lead inductance can be used to raise S_{22} to a large value, usually greater than unity even with a 50-ohm generator resistor. However, it is not necessary for the transistor S_{22} to be greater than unity, since the 50-ohm generator is not present in the oscillator design. The requirement for oscillation is $k < 1$; then resonating the input with a lossless termination will provide that $|S'_{22}| > 1$.

A simple example will clarify the design procedure. A common-base bipolar transistor (HP2001) was selected to design a fixed-tuned oscillator at 2 GHz. The common-base S-parameters and stability factor are given in Table 4.2. Using the load circuit in Fig. 4.7, we see that the input reflection coefficient is

$$S'_{11} = 1.18 \angle 173°$$

Thus a resonating capacitance of $C = 20$ pF resonates the input port. In a YIG-tuned oscillator, this reactive element could be provided by the high-Q YIG element.

4.4 Low-Noise Design

The design of low-noise oscillators is considerably more complicated than that of low-noise amplifiers, since the device usually operates in a nonlinear region where the device characteristics are difficult to measure. Before a discussion of low-noise oscillators can begin, it is best to start with a description of the common measurement methods and the results they yield. These differing measurement techniques give rise to noise descriptions that are related to one another. The following equations are common definitions of oscillator spectral purity:

$$S_\theta(f_m) = \text{spectral density of phase fluctuation}$$

$$S_\theta(f_m) = \Delta\theta^2_{\text{rms}} \tag{4.36}$$

$$S_{\dot\theta}(f_m) = \text{spectral density of frequency fluctuations}$$

$$S_{\dot\theta}(f_m) = \Delta f^2_{\text{rms}} \tag{4.37}$$

$$\mathcal{L}(f_m) = \text{the ratio of noise power in a 1-Hz bandwidth}$$
$$\text{at } f_m \text{ offset from carrier to carrier signal power}$$

$$\mathcal{L}(f_m) = \frac{N(1\text{--Hz BW})}{C} \tag{4.38}$$

The easiest technique for measuring oscillator noise is to view the oscillator spectrum directly on a spectrum analyzer, giving a display as in Fig. 4.8. This method allows direct measurement of $\mathcal{L}(f_m)$. The oscillator output power is read off the screen in dBm. The noise at a frequency offset f_m away from the carrier may also be read directly. Noise measured in this

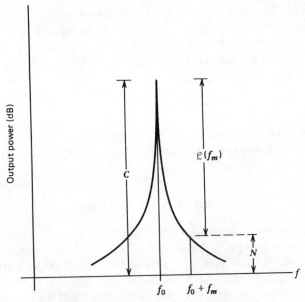

Figure 4.8   Oscillator output power spectrum.

way will usually require correction factors, since the detector of the analyzer is ordinarily an envelope rather than a true rms detector, the log amplifiers amplify noise peaks less, and the bandpass filters may be gaussian or trapezoidal in shape, which requires correction to a square bandpass. Additionally, since 1-Hz bandpass filters are uncommon, this results in measurement of the noise in a wider bandwidth which must be corrected to 1 Hz by reducing the noise measured by 10 dB for every decade by which the filter is wider than 1 Hz.

After applying these corrections, $\mathcal{L}(f_m)$ is equal to

$$\mathcal{L}(f_m) = \frac{\text{noise power with corrections at } f_m}{\text{carrier power}} = \frac{N(1-\text{Hz BW})}{C} \quad (4.39)$$

Certain precautions must be taken when measuring $\mathcal{L}(f_m)$ in this fashion. The technique is most useful when it can be determined that the noise of the oscillator being measured is worse than that of the local oscillator of the spectrum analyzer. The reason for this is apparent from Fig. 4.9a, which

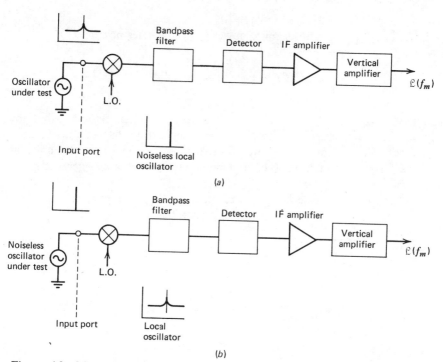

Figure 4.9 Measurement of noise-to-carrier ratio with spectrum analyzer. (a) Noiseless local oscillator; (b) Noiseless oscillator under test.

shows a spectrum analyzer from the front end. The noiseless local oscillator translates the oscillator under test to an IF frequency where the amplitude and noise can be analyzed with narrow fixed filters. The spectrum analyzer cannot distinguish between noise from its own local oscillator and that from an oscillator under test (which may be better) as in Fig. 4.9b. This situation frequently occurs at microwave frequencies where multiplied, low-noise oscillators often outperform the commonly used YIG-tuned oscillator in spectrum analyzers.

An important point that should be made is that the bulk of oscillator noise particularly close to the carrier is phase or FM noise. Oscillator limiting mechanisms, whether self-limiting or automatic gain control type, tend to eliminate AM noise. Under these conditions $\mathcal{L}(f_m)$ can be related to phase modulation in the following way. A table of Bessel functions will reveal that if a carrier is phase modulated (for a small modulation index $\Delta\theta_{\text{peak}} \ll \pi/2$), the ratio of the first-order sideband to the carrier J_0 is

$$\frac{J_1}{J_0} \simeq \frac{1}{2}\Delta\theta_{\text{peak}} \simeq \frac{1}{2}\sqrt{2}\,\theta_{\text{rms}} \tag{4.40}$$

Since $\mathcal{L}(f_m)$ is the ratio of noise power (J_1^2) to carrier power (J_0^2),

$$\mathcal{L}(f_m) = \frac{N}{C} = \left(\frac{J_1}{J_0}\right)^2 = \frac{1}{2}\theta_{\text{rms}}^2 \tag{4.41}$$

This description of $\mathcal{L}(f_m)$ holds only where it can be assumed that the $f_0 \pm \Delta f$ noise sidebands are correlated (i.e., caused by the same modulation source). This is not true in the additive noise region where the noise at $f_0 \pm \Delta f$ is not correlated (i.e., independent thermal noise generation at $\pm \Delta f$).

A more sensitive technique is to measure $S_\theta(f_m)$. Figure 4.10 shows a common oscillator noise measuring approach that gives $S_\theta(f_m)$.

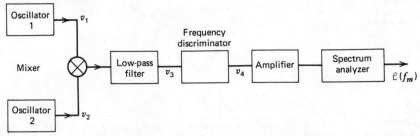

Figure 4.10 Measurement of spectral density of frequency fluctuations with frequency discriminator.

Describe oscillator 1 as

$$v_1 = V_1 \cos[\omega_1 t + \theta_1(t)] \tag{4.42}$$

and oscillator 2 as

$$v_2 = V_2 \cos[\omega_2 t + \theta_2(t)] \tag{4.43}$$

Then mixing these oscillators together gives an IF frequency low enough to apply to the frequency discriminator.

$$v_3 = V_1 V_2 \cos[\omega_1 t + d\theta_1(t)] \cos[\omega_2 t + d\theta_2(t)]$$

$$= \frac{V_1 V_2}{2} \cos\{(\omega_1 - \omega_2)t + [d\theta_1(t) - d\theta_2(t)]\} \tag{4.44}$$

The sum frequency term is eliminated by a low-pass filter. The output from the discriminator is

$$v_4 = K\omega_{in} \tag{4.45}$$

The frequency of the above signal is

$$\omega_4 = (\omega_1 - \omega_2) + \frac{d\theta_1(t) - d\theta_2(t)}{dt} \tag{4.46}$$

where $(\omega_1 - \omega_2)$ is a constant and $[d\theta_1(t) - d\theta_2(t)]/dt$ represents the sum of the frequency fluctuations in ω_1 and ω_2.

The output of the discriminator will be

$$v_4(t) = K\left((\omega_1 - \omega_2) + \frac{d\theta_1 - d\theta_2}{dt}\right) = K_2 + K\frac{d\theta_1 - d\theta_2}{dt} \tag{4.47}$$

A high-pass filter will remove the constant term, leaving

$$v_4(t) = K\frac{d\theta_1 - d\theta_2}{dt} = K\frac{d\theta_1}{dt} - K\frac{d\theta_2}{dt} = K \, d\omega_1 - K \, d\omega_2$$

$$= 2\pi K(df_1 - df_2) \tag{4.48}$$

This time function is then applied to a low-frequency spectrum analyzer. In the transform domain or frequency domain

$$v_4(f_m) = 2\pi K[dF_1(f_m) - dF_2(f_m)] \tag{4.49}$$

Since $dF_1(f_m)$ and $dF_2(f_m)$ are uncorrelated, they combine as follows:

$$. \Delta f_{\text{rms}} = dF_1(f_m) + dF_2(f_m) = \sqrt{\overline{(dF_1)^2} + \overline{(dF_2)^2}} \qquad (4.50)$$

$$v_4(f_m) = 2\pi K \sqrt{\overline{(dF_1)^2} + \overline{(dF_2)^2}} \qquad (4.51)$$

Since spectrum analyzers normally display power rather than voltage, the display represents

$$[v_4(f_m)]^2 = (2\pi K)^2 \left[\overline{(dF_1)^2} + \overline{(dF_2)^2} \right] \qquad (4.52)$$

It will be recognized that $\overline{dF_1^2} = S_{\dot\theta 1}(f_m)$ for oscillator one and $\overline{dF_2^2} = S_{\dot\theta 2}(f_m)$ for oscillator two.

The spectrum analyzer display is proportional to the sum of the spectral densities $S_{\dot\theta}(f_m)$ for oscillator one and oscillator two.

$$[v_4(f_m)]^2 = (2\pi K)^2 \left[S_{\dot\theta 1}(f_m) + S_{\dot\theta 2}(f_m) \right] \qquad (4.53)$$

$\mathcal{L}(f_m)$ now represents noise sideband power to carrier power caused by phase fluctuations as a function of frequency from the carrier. $S_{\dot\theta}(f_m)$ represents FM deviation squared as a function of frequency offset from the carrier.

The two parameters may be related as follows:

$$S_{\dot\theta}(f_m) = \overline{\Delta f(f_m)^2} \qquad (4.54)$$

$$df(t) = \frac{1}{2\pi} d\omega(t) = \frac{1}{2\pi} \frac{d\theta(t)}{dt} \qquad (4.55)$$

Then in the transform domain

$$df(f_m) = \frac{1}{2\pi}(s) d\theta(f_m) \qquad (4.56)$$

$$S_{\dot\theta}(f_m) = \overline{df(f_m)^2} = \left(\frac{s}{2\pi}\right)^2 \overline{d\theta(f_m)^2} \qquad (4.57)$$

$$\overline{d\theta(f_m)^2} = \left(\frac{2\pi}{s}\right)^2 S_{\dot\theta}(f_m) = \frac{1}{f_m^2} S_{\dot\theta}(f_m) \qquad (4.58)$$

$$\mathcal{L}(f_m) = \frac{1}{2} \overline{\theta(f_m)^2} = \frac{1}{2f_m^2} S_{\dot\theta}(f_m) \qquad (4.59)$$

This technique affords better sensitivity than direct measurement of $\mathcal{L}(f_m)$ at microwave frequencies, since the translation down to low rf frequencies permits the use of spectrum analyzers with lower-noise local oscillators or fast Fourier transform analyzers. In general, the sensitivity of this system is limited by the internal noise of the frequency discriminator.

A more sensitive scheme removes the frequency discriminator as shown in Fig. 4.11. We assume that the oscillators are or can be adapted so that one can be phase-locked to the other. In Fig. 4.11 the oscillators are set so that they are at approximately the same frequency. Oscillator one and oscillator two then mix to produce sum and difference frequencies. The sum frequencies are removed by the low-pass filter. The difference frequency error signal is sent back to lock oscillator two to oscillator one. Inside the loop bandwidth, which can be adjusted by varying the gain of G_1, the noise of oscillator two tracks that of oscillator one. Outside the loop bandwidth, the noise of the two oscillators shows no correlation.

The mixer is usually a double-balanced mixer consisting of four diodes. The IF port is dc-coupled to provide the phase-lock dc signal. This phase-lock dc signal is adjusted to be 0 V on the voltmeter, since the sensitivity $dv/d\theta$ is maximum for this condition. This is done by adjusting a line length such that the phases of the two oscillators are 90° apart.

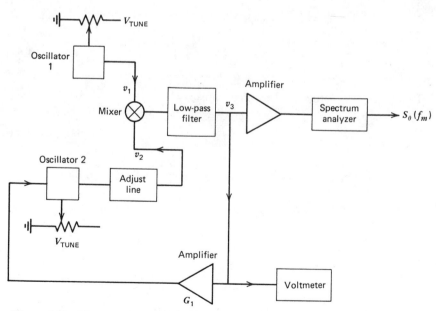

Figure 4.11 Measurement of spectral density of phase fluctuations with a mixer.

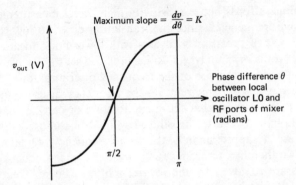

Figure 4.12 Output voltage of doubly balanced mixer versus phase difference between local oscillator and rf signal ports.

Figure 4.12 shows the typical sensitivity of the mixer. Beyond the loop bandwidth, the output of the mixer may be described as follows:

$$v_1 = V_1 \cos(\omega t + \theta_{n1}) \tag{4.60}$$

$$v_2 = V_2 \cos(\omega t + \theta_{n2} - \pi/2) \tag{4.61}$$

$$v_3 = V_1 V_2 \cos(\omega t + \theta_{n1}) \cos(\omega t + \theta_{n2} - \pi/2)$$

$$= \frac{V_1 V_2}{2} \cos\left(\theta_{n1} - \theta_{n2} + \frac{\pi}{2}\right) \tag{4.62}$$

The θ_{n1} and θ_{n2} terms are rms phase noise, which can be combined as

$$\theta_{nT} = \sqrt{\overline{\theta_{n1}^2} + \overline{\theta_{n2}^2}} \tag{4.63}$$

$$v_3 = \frac{V_1 V_2}{2} \cos\left(\theta_{n1} - \theta_{n2} + \frac{\pi}{2}\right) = \frac{V_1 V_2}{2} \cos\left(\theta_{nT} + \frac{\pi}{2}\right)$$

$$= -\frac{V_1 V_2}{2} \sin(\theta_{nT}) \tag{4.64}$$

For θ_{nT} very small,

$$\sin(\theta_{nT}) \simeq \theta_{nT} = \sqrt{\overline{\theta_{n1}^2} + \overline{\theta_{n2}^2}} \tag{4.65}$$

Since the spectrum analyzer displays power, it will show the square of the term

$$-\frac{V_1 V_2}{2}\sqrt{\overline{\theta_{n1}^2 + \theta_{n2}^2}} \quad \text{or} \quad \left(\frac{V_1 V_2}{2}\right)^2 \left(\overline{\theta_{n1}^2} + \overline{\theta_{n2}^2}\right)$$

$$\overline{\theta_{n1}^2} = S_\theta(f_m) \text{ of oscillator one} \tag{4.66}$$

$$\overline{\theta_{n2}^2} = S_\theta(f_m) \text{ of oscillator two} \tag{4.67}$$

If the spectral densities have equal power distribution but are not correlated, then the mixer output is 3 dB greater than either one alone. This
technique yields the sum of the $S_\theta(f_m)$ for oscillator one and oscillator two.

$S_\theta(f_m)$ can now be related to $\mathcal{L}(f_m)$. $S_\theta(f_m)$ is equal to $\mathcal{L}(f_m)$ folded
about itself. Therefore, $S_\theta(f_m) = 2\mathcal{L}(f_m)$ if the noise sidebands about f_1 are
correlated and $S_\theta(f_m) = \sqrt{2}\,\mathcal{L}(f_m)$ if they are not correlated.

In Fig. 4.13 the noise below $f_0 - \Delta f$ is assumed to be uncorrelated to the
noise above $f_0 + \Delta f$ in oscillator one and in oscillator two. Closer than

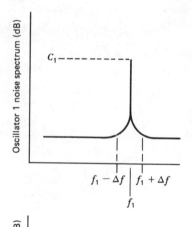

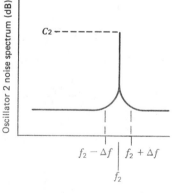

Figure 4.13 Noise spectrum of two oscillators
at f_1 and f_2 carrier frequency.

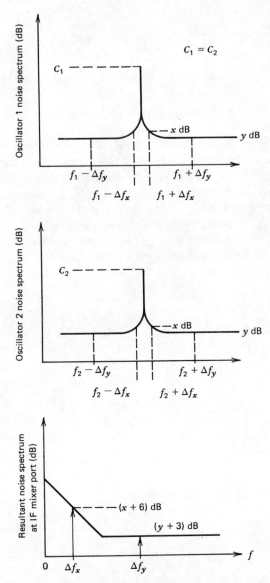

Figure 4.14 Resultant noise spectrum due to foldover of spectrum about the carrier for one oscillator.

154

$f_0 \pm \Delta f$ the assumption is that there is correlation of the noise above and below the carrier in both oscillators. Beyond $\pm \Delta f$ we assume that this is the noise floor of the device. Closer than $\pm \Delta f$ we assume that the noise is caused by phase modulation mechanisms in the device or other components that generate related sidebands above and below the carrier. When these two spectrums are mixed together, the following occurs: If $f_1 = f_2$, then (if we ignore the sum frequency components, which are eliminated by the low-pass filter), $f_1 - f_2 = 0$; f_1 then mixes against the noise spectrum of f_2. This causes the noise spectrum of f_2 to fold upon itself. For instance, f_1 mixing against $f_2 \pm \Delta f_x$ will yield two correlated noise components at Δf_x which add in power to cause a 6-dB increase as in Fig. 4.14. However, if f_1 mixes against $f_2 \pm \Delta f_y$, there is only a 3-dB increase, since the noise at $f_2 - \Delta f_y$ is not correlated to that at $f_2 + \Delta f_y$.

The reverse also occurs: f_2 can mix with the noise of f_1 at $f_1 \pm \Delta f_x$ and $f_1 \pm \Delta f_y$ to cause an additional 3-dB increase in noise measured at the mixer's output. This increase occurs because this reverse process generates another spectrum identical in amplitude to that in Fig. 4.14; however, the noise of the two oscillators is not correlated except within the phase-locked loop bandwidth. The mixer takes these two uncorrelated spectrums and adds them at its output, causing an additional 3-dB increase in noise, as shown in Fig. 4.15. $\mathcal{L}(f_m)$ can be obtained from this spectrum by subtracting 9 dB from the part where the upper and lower noise sidebands are correlated and by subtracting 6 dB from the area where no correlation exists.

It is possible to go back to Fig. 4.14 before the addition of 3 dB (due to two uncorrelated oscillators) to see how $S_\theta(f_m) = \Delta\theta^2_{\text{rms}}$ is related to $\mathcal{L}(f_m)$. Since

$$\mathcal{L}(f) = \frac{1}{2} \Delta\theta^2_{\text{rms}} \tag{4.68}$$

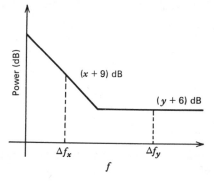

Figure 4.15 Power spectrum of mixer IF port as displayed on a spectrum analyzer due to the combined effects of foldover and addition of 3 dB for noise spectrum of two uncorrelated oscillators.

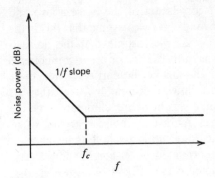

Figure 4.16 Noise power versus frequency of a transistor amplifier.

and

$$S_\theta(f_m) = \Delta\theta_{\text{rms}}^2 \qquad (4.69)$$

we see that the folded-over spectrum of a single oscillator at Δf_x or where the upper and lower sidebands of f_1 are correlated is equal to $S_\theta(f_m)$.

The noise spectrum of an amplifier would appear as in Fig. 4.16. For a moment, it is of interest to discuss the $1/f$ noise spectrum near dc. Noise in amplifiers is often modeled as in Fig. 4.17. In bipolar amplifiers, e_n is related to the thermal noise of the base spreading resistance:

$$e_n = \sqrt{4kTr_bB} \qquad (4.70)$$

This noise source has a relatively flat frequency response. The i_n noise source is associated with the shot noise in the base current:

$$i_n = \sqrt{2qI_bB} \qquad (4.71)$$

This i_n noise generator has associated with it a $1/f$ noise mechanism.

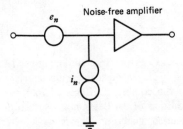

Figure 4.17 Equivalent noise sources at the input of an amplifier.

In FET devices, the situation is reversed. The e_n noise generator has a $1/f$ noise component where i_n shows none. It is interesting to note that in general, the $1/f$ noise corner of bipolar silicon devices is lower than that of silicon JFETs. Silicon JFETs are less noisy than silicon MOSFETS. GaAs MESFETs usually have the highest $1/f$ corner frequencies, which can extend to several hundred megahertz. Carefully selected bipolar devices can have $1/f$ noise corners below 100 Hz.

There are various instruments that can measure e_n and i_n directly with no carrier signal present. These measurement methods would provide a noise plot as in Fig. 4.16. However, if a carrier signal is applied to the amplifier, the noise plot would be modified as in Fig. 4.18. The low-frequency noise sources can effect the phase shift through the amplifier causing the $1/f$ phase noise spectrum about the carrier.

Since an oscillator can be viewed as an amplifier with feedback, it is helpful to examine the phase noise added to an amplifier that has a noise figure F. With F defined by

$$F = \frac{(S/N)_{\text{in}}}{(S/N)_{\text{out}}} = \frac{N_{\text{out}}}{N_{\text{in}}G} = \frac{N_{\text{out}}}{GkTB} \qquad (4.72)$$

$$N_{\text{out}} = FGkTB \qquad (4.73)$$

$$N_{\text{in}} = kTB \qquad (4.74)$$

where N_{in} is the total input noise power to a noise-free amplifier. The input phase noise in a 1-Hz bandwidth at any frequency $f_0 + f_m$ from the carrier

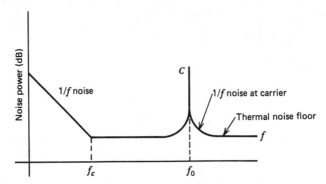

Figure 4.18 Noise power versus frequency of a transistor amplifier with an input signal applied.

produces a phase deviation given by (Fig. 4.19).

$$\Delta\theta_{\text{peak}} = \frac{V_{\text{nRMS1}}}{V_{\text{avsRMS}}} = \sqrt{\frac{FkT}{P_{\text{avs}}}} \tag{4.75}$$

$$\Delta\theta_{1\,\text{RMS}} = \frac{1}{\sqrt{2}}\sqrt{\frac{FkT}{P_{\text{avs}}}} \tag{4.76}$$

Since a correlated random phase relation exists at $f_0 - f_m$, the total phase

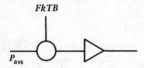

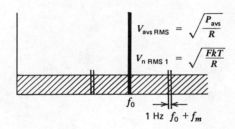

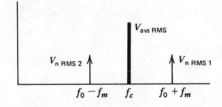

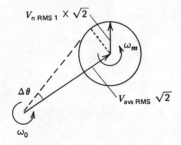

Figure 4.19   Phase noise added to carrier.

deviation becomes

$$\Delta\theta_{\text{RMS total}} = \sqrt{\frac{FkT}{P_{\text{avs}}}} \tag{4.77}$$

The spectral density of phase noise becomes

$$S_\theta(f_m) = \Delta\theta^2_{\text{RMS}} = \frac{FkTB}{P_{\text{avs}}} \tag{4.78}$$

where $B = 1$ for a 1-Hz bandwidth.
Using

$$kTB = -174 \text{ dBm/Hz} \tag{4.79}$$

allows a calculation of the spectral density of phase noise that is far removed from the carrier i.e., at large values of f_m. This noise is the theoretical noise floor of the amplifier. For example, an amplifier with $+10$ dBm power at the input and a noise figure of 6 dB gives

$$S_\theta(f_m > f_c) = -174 \text{ dBm} + 6 \text{ dB} - 10 \text{ dBm} = -178 \text{ dB}$$

For a modulation frequency close to the carrier, $S_\theta(f_m)$ shows a flicker or $1/f$ component which is empirically described by the corner frequency f_c. The phase noise can be modeled by a noise-free amplifier and a phase modulator at the input as shown in Fig. 4.20. The purity of the signal is degraded by the flicker noise at frequencies close to the carrier. The spectral

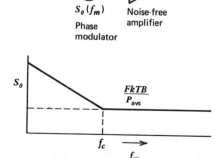

Figure 4.20 Phase noise modeled by a noise-free amplifier and a phase modulator.

phase noise can be described by

$$S_\theta(f_m) = \frac{FkTB}{P_{avs}}\left(1 + \frac{f_c}{f_m}\right) \tag{4.80}$$

The oscillator may be modeled as an amplifier with feedback as shown in Fig. 4.21. The phase noise at the input of the amplifier is affected by the bandwidth of the resonator in the oscillator circuit in the following way. The tank circuit or bandpass resonator has a low-pass transfer function

$$L(\omega_m) = \frac{1}{1 + j(2Q_L\omega_m/\omega_0)} \tag{4.81}$$

where

$$\frac{\omega_0}{2Q_L} = \frac{B}{2} \tag{4.82}$$

is the half bandwidth of the resonator. These equations describe the

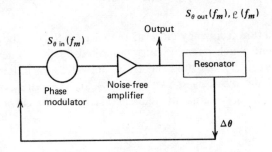

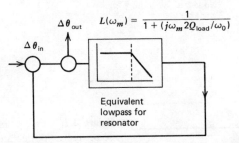

Figure 4.21   Equivalent feedback models of oscillator phase noise.

amplitude response of the bandpass resonator; the phase noise is transferred unattenuated through the resonator up to the half bandwidth. The closed loop response of the phase feedback loop is given by

$$\Delta\theta_{out}(f_m) = \left(1 + \frac{\omega_0}{j2Q_L\omega_m}\right)\Delta\theta_{in}(f_m) \tag{4.83}$$

The power transfer becomes the phase spectral density

$$S_{\theta out}(f_m) = \left[1 + \frac{1}{f_m^2}\left(\frac{f_0}{2Q_L}\right)^2\right]S_{\theta in}(f_m) \tag{4.84}$$

where $S_{\theta in}$ was given by (4.80). Finally, $\mathcal{L}(f_m)$ is

$$\mathcal{L}(f_m) = \frac{1}{2}\left[1 + \frac{1}{f_m^2}\left(\frac{f_0}{2Q_L}\right)^2\right]S_{\theta in}(f_m) \tag{4.85}$$

This equation describes the phase noise at the output of the amplifier. The phase pertubation $S_{\theta in}$ at the input of the amplifier is enhanced by the positive phase feedback within the half bandwidth of the resonator, $f_0/2Q_L$.

Depending on the relation between f_c and $f_0/2Q_L$, there are two cases of interest as shown in Fig. 4.22. For the low-Q case, the spectral phase noise is unaffected by the Q of the resonator, but the $\mathcal{L}(f_m)$ spectral density will show a $1/f^3$ and $1/f^2$ dependence close to the carrier. For the high-Q case, a region of $1/f^3$ and $1/f$ should be observed near the carrier. Substituting (4.80) in (4.85) gives an overall noise of

$$\mathcal{L}(f_m) = \frac{1}{2}\left[1 + \frac{1}{f_m^2}\left(\frac{f}{2Q_L}\right)^2\right]\frac{FkTB}{P_{avs}}\left(1 + \frac{f_c}{f_m}\right)$$

$$= \frac{FkTB}{2P_{avs}}\left[\frac{1}{f_m^3}\frac{f^2f_c}{2Q_L} + \frac{1}{f_m^2}\left(\frac{f}{2Q_L}\right)^2 + \frac{f_c}{f_m} + 1\right] \tag{4.86}$$

Examining (4.86) gives the four major causes of oscillator noise: the up-converted $1/f$ noise or flicker FM noise, the thermal FM noise, the flicker phase noise, and the thermal noise floor, respectively.

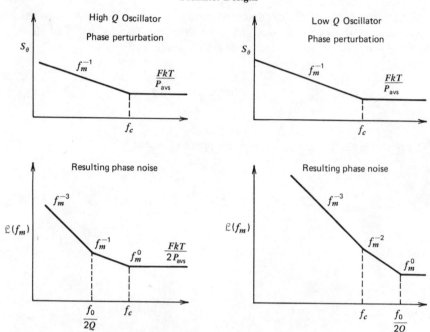

Figure 4.22 Oscillator phase noise for high-Q and low-Q resonator viewed as spectral phase noise and as noise-to-carrier ratio versus frequency from the carrier.

4.5   High-Power Design

High-power oscillators refer to circuits in which a transistor is operating near its maximum output power rating. There are two types of high-power oscillators. The simplest design uses a single high-power transistor as the oscillator device. This is usually a high-power amplifier transistor that can be used for either application. The other type is a low-power oscillator followed by a high-power amplifier (which could include several cascaded transistors). The amplifier is referred to as a buffer, and the design details for buffered oscillators are found in Section 4.7. The advantage of the buffered design is an improved output noise spectrum and reduced frequency pulling when the load impedance changes. The disadvantages are higher cost, larger size, and lower dc-rf efficiency.

The design of high-power amplifiers and oscillators is very similar. The same transistors are used, the same dc bias points are used, and the same S-parameters are used. The large-signal S-parameters should be used if they are available. The output port is delivering nearly the maximum power possible for both designs. Since the output load is accepting power from a connector, the circuit could be either an amplifier or an oscillator.

For the case of a high-power oscillator, the transistor must feed back a significant portion of its available output power to the input port. This might be done by an external feedback element or internally within the transistor. The net gain in the forward and feedback loop is unity, but the transducer gain or S_{21} of the transistor is normally greater than unity. Since some of the available output power is required at the input to sustain oscillations, the oscillator output power is slightly lower than the output power when used as an amplifier. The maximum value of $P_{LOAD} - P_{in}$ is the same for both applications.

Two additional design criteria are required for the oscillator design. Since the transistor is operating under a large-signal condition, possibly in a Class C bias condition, the dynamic load line may be dangerously close to the unsafe operating range where permanent damage can occur. This can be avoided by observing the output spectrum as the dc bias is increased. If the noise increases dramatically at full bias, the transistor may be on the verge of burnout, the overall design is marginal, and a higher-power transistor should be used.

The other precaution with high-power oscillators is to insure that the dc biasing is included in the design. Input and output short-circuited transmission lines are useful for this function. A resistive rf choke is not recommended because of reduced efficiency. The bias lines must be properly filtered for low frequencies to avoid unwanted low-frequency oscillations or "motor-boating."

Another consideration in high-power design is the possibility that the resonator will change its properties at high power levels. An example of this is a YIG, which begins to limit at relatively low powers. As the YIG begins to limit, the loaded Q reduces and the noise of the oscillator increases. For this reason, it is not possible to operate YIG oscillators at very high power levels without the use of buffer amplifiers. Another example is a varactor-tuned oscillator. If the power level at the varactor is very high, harmonic generation by the varactor will limit the nonspurious response of the oscillator. For lowest harmonic distortion, the oscillator transistor is normally biased at the point for maximum Class A output power.

An example of a high-power GaAs MESFET oscillator design is shown in Figs. 4.23 and 4.24. The transistor was mounted in a stripline package and characterized with large-signal S-parameters in the common-gate configuration. These data were derived by conjugately matching the transistor at the desired operating signal level and measuring the tuner impedances. The conjugate of the input and output tuner impedances were used as the S_{11} and S_{22} data. Small-signal S_{21} and S_{12} were measured and adjusted to large-signal values by lowering S_{21} to agree with the measured value of large-signal transducer power gain [see (1.142)].

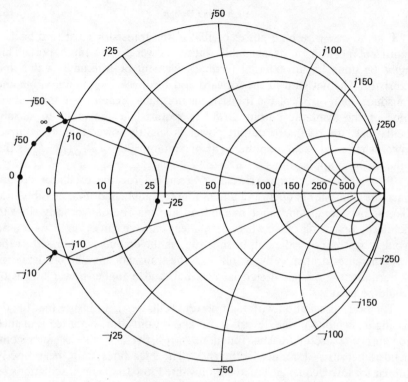

Figure 4.23 Output drain impedance S'_{22} versus input source impedance Γ_G. Coordinates in ohms. DXL4640-P100; $V_{DS} = 8.0$ V; $I_{DS} = 450$ mA; $f = 8$ GHz.

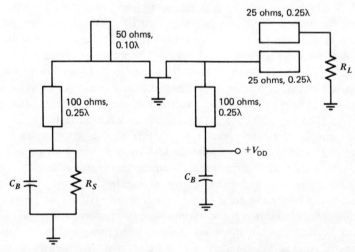

Figure 4.24 High-power oscillator circuit at 8 to 10 GHz. DXL4640A-P100 1 W, 8 GHz; DXL4615A-P100 0.25 W, 10 GHz; $V_{DS} = 10$ V; $I_{DS} = 0.5 I_{DSS}$.

The oscillator structure is the series-resonant structure with power out the drain (Fig. 4.3). The effect of lossless termination Γ_G on the input port is plotted in the S_{22} plane at 8 GHz in Fig. 4.23. This map shows the output port unstable for all values of inductive termination and most values of capacitance termination. For $X_G = -j25$ ohms, the output port becomes stable. Also, the output port is nearly self-resonant for any lossless input termination. The required value of load impedance for $\Gamma_L S_{22}' = 1$ is about 12 ohms, which can be achieved by a quarter-wave transformer of 25 ohms. This transformer is realized by a quarter-wavelength parallel-coupled line adjusted for critical coupling. The complete oscillator schematic including dc biasing is shown in Fig. 4.24. The oscillating frequency can be adjusted by the length of the input stub and the length of the critically coupled transformer.

The resulting oscillator gives an output power of 1 W at 7 to 8 GHz for the DXL4640A-P100 transistor and 0.25 W at 8 to 10 GHz for the DXL4615A-P100 transistor. This is in good agreement with the amplifier output power in the common-source mode with a large-signal gain of 3 to 5 dB.

The thermal impedance for this design is estimated as

$$\text{DXL4640A-P100} \quad \theta_j \simeq 25°C/W$$

$$\text{DXL4615A-P100} \quad \theta_j \simeq 40°C/W$$

The junction temperature is

$$\Delta T = \theta_j (P_{dc} - P_{LOAD})$$

$$\text{DXL4640A} \quad \Delta T = 25(3.6 - 1) = 65°C$$

$$\text{DXL4615A} \quad \Delta T = 40(1.8 - 0.25) = 62°C \tag{4.87}$$

At an ambient temperature of 25°C, the junction is operating at about 90°C, a safe operating temperature for long-term reliability. The design procedure for high-power oscillators can be summarized as follows:

1 Select a power transistor capable of the required output power at the frequency of operation. Usually the transistor can be used for either an amplifier or an oscillator application.

2 Select the bias point from the recommended point for maximum Class A output power.

3 From small-signal S-parameters, determine the resonating reactance at the input port.

4　Select the load circuit which is required for $\Gamma_L S'_{22} = 1$. Usually the amplifier and oscillator output matching structure are nearly the same, since the load cannot distinguish between amplifier and oscillator.

5　Check that the resonator does not saturate at the operating power level.

6　Calculate the junction temperature at the dc operating point.

After the oscillator is fabricated, the harmonics and output noise spectrum must be evaluated. If the harmonics are high, increase the dc operating bias or find a larger transistor. If the output noise spectrum is too high, consider a higher-Q resonator or a buffered oscillator design.

4.6　Broadband Design

For broadband oscillator design, the stability factor must be less than unity for the entire frequency range and the resonator must be electrically or mechanically tunable. The most common broadband resonators are the YIG and the hyperabrupt varactor. Special consideration must be given to the load circuit to guarantee oscillation over the entire operating range. This is accomplished by plotting the stability circles in the output plane or Γ_L plane. The Γ_L presented to the transistor output must fall in the unstable region over the required bandwidth.

One technique for broadband oscillator design is to design the load circuit for a nearly constant value of $S'_{11} \simeq 1.2$. This guarantees oscillation when resonated at the input port. In addition, because of (4.33), the transducer power gain or $|S_{21}|$ of the transistor should be nearly constant over the operating frequency range. It has been experimentally verified that a rapid decrease in $|S_{21}|$ with frequency corresponds to a drop in output power with frequency. This is predicted by (4.33).

The requirement for constant $|S_{21}|$ for the oscillator is only needed if constant output power is needed. Another method of achieving this is to add a buffer amplifier that has a sloped $|S_{21}|$ to give an overall constant transducer gain. For the buffered design, both the interstage and output networks may be adjusted for a constant transducer gain. The broadband buffered design requires both $S'_{11} \simeq 1.2$ and a constant $|S_{21}|$ for the two stages. The gain-bandwidth limitations presented in Figs. 3.12 and 3.13 apply to the matching structures in the broadband buffered design.

The speed of tuning the resonator is often an important design limitation. The YIG resonator is relatively slow, usually requiring milliseconds to sweep an octave and return to the starting frequency. The varactor tunes much faster (usually in nanoseconds), but it reaches steady state more slowly, which is the cause of posttuning drift in varactor oscillators. The causes of posttuning drift can be either thermal effects in the varactor or surface states in the varactor. This can cause an additional settling period of 10 to 100 nsec.

The bandwidth limitation for the oscillator is not the transistor. Usually the bandwidth limitation factor is the feedback or the resonator. Since the transistor should oscillate up to f_{max}, in principle it can be made to oscillate over many octaves below f_{max}; for example, 2 to 18 GHz oscillators should be possible. The difficulty in this design is keeping the stability factor less than unity and providing a resonator with such a wide bandwidth. As circuit techniques improve, broader bandwidth oscillators should be developed. The transistors are not the inherent limitation.

4.7 Buffered Oscillator Design

The buffered oscillator consists of an oscillator transistor operating into an amplifier transistor as shown in Fig. 4.25. This allows the oscillator to be optimized at lower power levels, where the noise spectrum can be minimized. The amplifier raises the power level and reduces the effect of load variations on the oscillator frequency. The amplifier buffers the load from the input resonator.

The design procedure for a buffered oscillator may be summarized as follows:

1 Check the stability factor of the oscillator transistor over required frequency range. Add feedback if required to give $k < 1$ over the entire oscillator frequency range.

2 With the amplifier transistor as the load (assuming that $S_{12B} \simeq 0$ for simplicity) design M_2 so that the input port of the oscillator gives $S'_{11A} > 1$ over the entire oscillator frequency range.

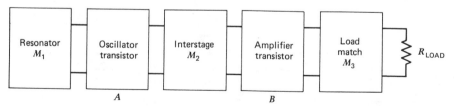

Figure 4.25 Buffered oscillator.

3 Resonate the oscillator M_1 port so that

$$\Gamma_S S'_{11A} = 1$$

$$\Gamma_L S'_{22A} = 1$$

over the entire frequency range, which requires a resonator that can be tuned either electrically or mechanically.

4 Convert the oscillator transistor to an equivalent generator circuit and design amplifier circuits M_2 and M_3 for constant or sloped gain, depending on the oscillator output power. This step is shown in Fig. 4.26.

A common procedure for buffered oscillator design is:

1 Design an oscillator circuit with a 50-ohm load.
2 Design an amplifier circuit with a 50-ohm generator and a 50-ohm load.
3 Connect the oscillator circuit to the amplifier circuit.

This procedure will work only if the input to the amplifier is exactly 50 ohms. If the buffered oscillator is wideband, this is usually not possible. This procedure is analogous to a two-stage amplifier design where each stage is matched to a 50-ohm point and the two stages are cascaded. In the two-stage amplifier design, it is recommended to design the interstage from

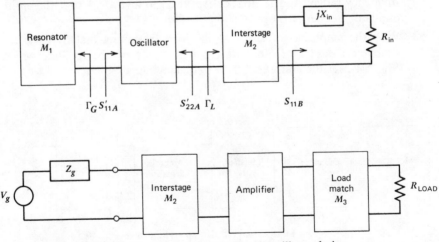

Figure 4.26 Two steps in buffered oscillator design.

the output of the first stage directly into the input of the second stage without regard to a reference 50-ohm point. The same principle should be applied to buffered oscillator design for best success over a wide bandwidth.

The conversion of the oscillator to an equivalent source impedance has been outlined in Fig. 4.27. For constant power into the load over frequency, the designer must examine I_0 and R_L versus frequency. If these parameters decrease with frequency, the input power to the amplifier is decreasing with frequency. For the wideband designs, it is helpful to maintain R_L and I_0 nearly constant with frequency. If R_L increases with frequency, the buffer amplifier may have a gain rolloff; however, the oscillator will probably not oscillate at the high-frequency end of the band.

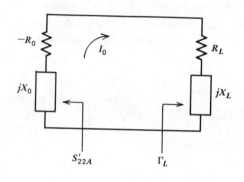

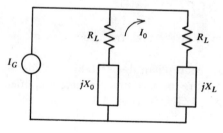

$$-R_0 = R_L$$

$$-X_0 = X_L$$

$$P_{avs} = I_0^2 R_L = I_0^2 |R_0|$$

$$P_{avs} = I_0^2 R_L = \frac{E_G^2}{4R_L}$$

$$E_G = 2I_0 R_L$$

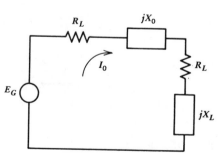

Figure 4.27 Equivalent source impedance of oscillator.

The available power from the oscillator is the power delivered to the input of the buffer amplifier. This power level can be estimated from (4.33) and Fig. 4.27. The conditions given by (4.2) and (4.3) require that a conjugate match be given at the output of the oscillator. Since the gain G may decrease with frequency, (4.33) predicts that the oscillator power normally decreases with frequency. Thus, R_L will probably decrease with frequency, requiring gain sloping in the buffer amplifier for constant output power. This gain sloping is achieved by proper design of M_2 and M_3, with about equal gain slopes in these two networks. Since a conjugate match exists at the input of M_2 over the entire oscillator frequency range, a frequency selective mismatch at the input of the transistor requires a resistor in M_2. The alternative approach is placing all of the mismatch in M_3.

The buffer amplifier problem can also be treated by converting the oscillator to an equivalent generator b_G as described in Section 1.4. In this case, the amplifier problem is given in Fig. 4.28. The transducer gain of this amplifier is, from (1.141),

$$G_{Tu} = \frac{P_{out}}{P_{avs}} = \frac{\left(1 - |\Gamma_2|^2\right)|S_{21B}|^2\left(1 - |\Gamma_1|^2\right)}{|(1 - S_{22B}\Gamma_2)|^2|(1 - S_{11B}\Gamma_1)|^2} \tag{4.88}$$

which reduces for a simultaneous conjugate match to

$$G_{Tumax} = \frac{|S_{21B}|^2}{\left(1 - |S_{22B}|^2\right)\left(1 - |S_{11B}|^2\right)} \tag{4.89}$$

This is the gain that should be achieved at the high-frequency end of the band.

A design example will illustrate the concepts in this section. Consider a common-gate MESFET as an oscillator driving a common-source amplifier MESFET. The S-parameters are given at 4 GHz in Table 4.3.

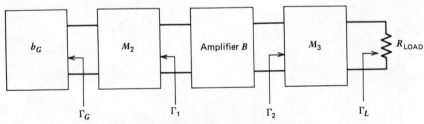

Figure 4.28 Buffer amplifier design.

Table 4.3   S-Parameters of DXL2501A GaAs MESFET Chip

CS	CG	
$S_{11} = 0.83 \angle -67°$	$S_{11} = 0.20 \angle -180°$	$V_{DS} = 6.0$ V
$S_{21} = 2.16 \angle 119°$	$S_{21} = 1.2 \angle -24°$	$I_{DS} = 30$ mA
$S_{12} = 0.07 \angle 61°$	$S_{12} = 0.146 \angle 25°$	$V_{GS} = -3$ V
$S_{22} = 0.66 \angle -23°$	$S_{22} = 0.92 \angle -14°$	
$k = 0.66$	$k = 0.69$	$f = 4$ GHz

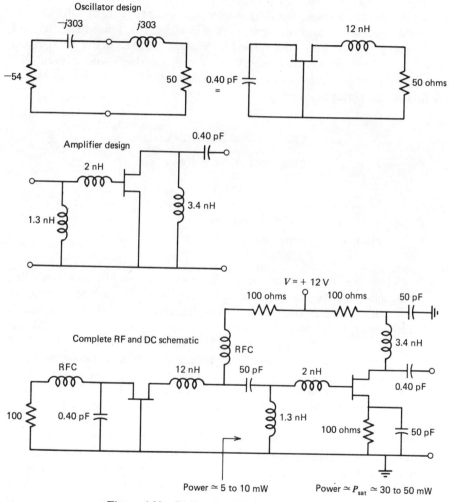

Figure 4.29   Buffered oscillator at 4 GHz.

171

Calculating S'_{22} for the CG device with a passive termination of $-j100$ ohms (0.40 pF) gives

$$S'_{22} = 1.06 \angle -18°$$

$$Z_{out} = -54 - j303 \text{ ohms}$$

The required load impedance is $50 + j303$ ohms, which is a 12-nH inductor in series with 50 ohms. The oscillator circuit is given in Fig. 4.29. The amplifier can also be designed using Section 3.1 with $S_{12} = 0$. The maximum overall gain is 14.9 dB, and the complete rf circuit is given in Fig. 4.29 with the addition of dc bias. The output transistor will probably be saturated for this design, since the small-signal gain is so large. The active bias circuit of Section 2.3 could be used for the buffer amplifier. The active bias should not be used on the oscillator transistor, since this could prevent the transistor from oscillating or produce low-frequency noise due to oscillations in the active bias circuit.

4.8 Practical Oscillator Circuits

There are essentially three types of oscillator circuits:

1 Fixed-tuned oscillators.
2 Magnetically tuned YIG oscillators.
3 Voltage-tuned varactor oscillators.

In the fixed-tuned group, the tuning may be accomplished by coupling to a high-Q cavity resonator or to a reactive feedback circuit. For the fixed-tuned resonator circuits, the lowest-frequency mode of the resonator is usually chosen for stable operation. The most common reactive feedback oscillators are shown in Fig. 4.30, where each of these circuits is named for the original inventor of the oscillator feedback. To visualize the common-base (or common-gate) circuit operation, the collector and emitter currents have nearly zero phase shift at low frequency. Thus a tank circuit with nearly zero phase shift will provide the correct phase relationship for oscillation. For the common-emitter and common-collector circuit operation, the input and output currents are nearly 180° out of phase at low frequency. The tank circuit must provide another 180° of phase shift to give the resonance condition. The inductive feedback circuits are the Armstrong and Hartley oscillators; the capacitive feedback circuits are the Colpitts and Clapp oscillators. The Colpitts and Hartley circuits have been redrawn in Fig. 4.31 to aid the recognition of these oscillators.

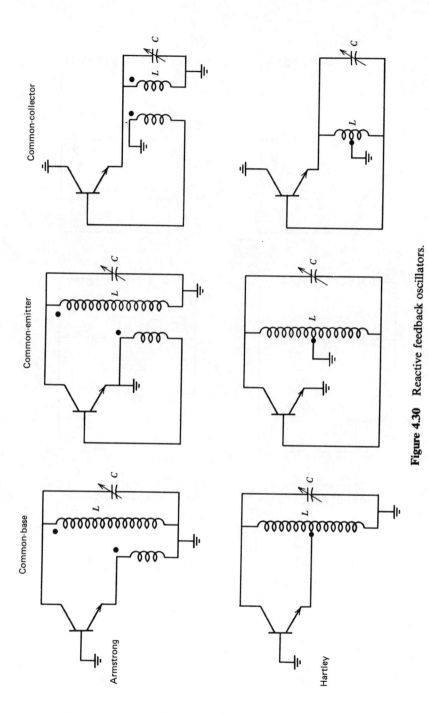

Figure 4.30 Reactive feedback oscillators.

173

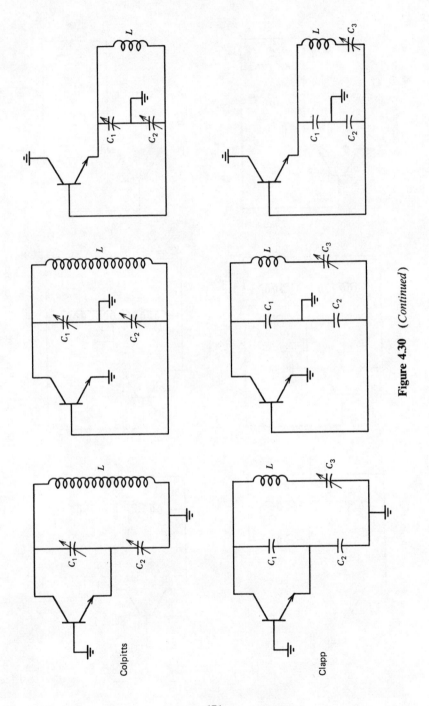

Colpitts

Clapp

Figure 4.30 (*Continued*)

174

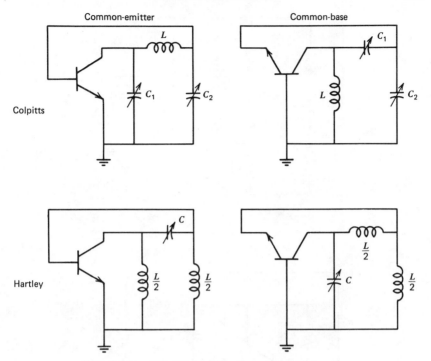

Figure 4.31 Colpitts and Hartley oscillator circuits.

Since ground is simply a reference node in the circuit, the choice of the equivalent ac circuit for analysis may be arbitrary. For example, if load power is taken from the collector (or drain) lead, the oscillator may be either common-base or common-emitter, depending on the location of the ground terminal. Thus an exact definition of oscillator type may not always be possible. The choice of the ground node may be due to practical considerations such as dc biasing or easy adjustment of the feedback network.

An example of two very similar Colpitts oscillators is shown in Fig. 4.32. The dc biasing circuit is common-emitter for both cases. The large bypass capacitor across R_1 effectively grounds the base for the common-base oscillator. For the common-emitter oscillator, the blocking capacitor is required to prevent dc shorting of the collector to base. For both oscillators, the collector is loaded by C_B and R_L.

Another example of a fixed-tuned oscillator is shown in Fig. 4.33, where a directional coupler and an amplifier are combined so that the net gain in the feedback loop is unity. The net phase shift in the feedback loop must be a multiple of 2π at the resonant frequency. Often a dielectric resonator is

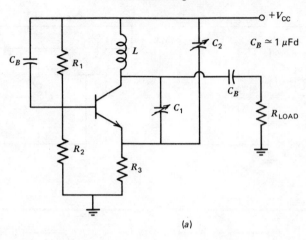

(a)

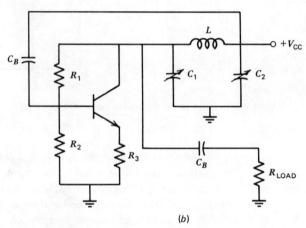

(b)

Figure 4.32 Two Colpitts oscillators. (a) Common-base Colpitts oscillator; (b) Common-emitter Colpitts oscillator.

added in the feedback loop to provide a rapid phase shift at the desired resonant frequency.

The magnetically tuned YIG oscillator is shown in Fig. 4.34. Since the output is normally at the collector, the oscillator may be analyzed either common-emitter or common-base. Common-base is normally chosen, since the feedback is easily visualized in this circuit. Increasing the base inductance will make the device conditionally stable at lower frequencies. The design of broadband YIG oscillators using small-signal S-parameters is

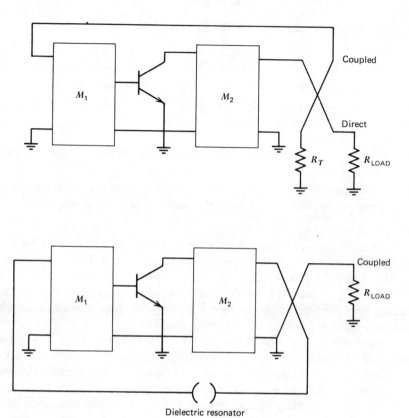

Figure 4.33 Fixed-tuned oscillators using amplifier and directional coupler.

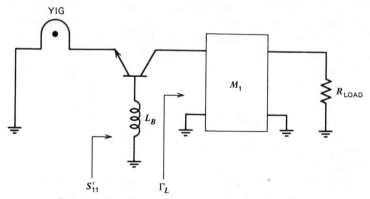

Figure 4.34 Common-base YIG-tuned oscillator.

straightforward. A recommended design procedure follows:

1 Select the base inductance for $k < 1$ over the desired frequency range.
2 Model the YIG element as a high-Q, parallel-resonant structure.
3 At each frequency find the load impedance Γ_L which allows oscillation (i.e., $S'_{22}\Gamma_L = 1$). This defines a region in the Smith Chart plane that permits oscillation.
4 Synthesize the M_1 network to provide the desired Γ_L over the frequency range of oscillation.
5 Simultaneously require $|S_{21}|$ to be constant with frequency for constant output power. This may require the addition of a buffer amplifier.

The impedance presented at the emitter is usually inductive, so the YIG presents a capacitance to resonate the oscillator. If the impedance presented to the YIG becomes capacitive, the transistor may resonate with the parasitic loop inductance for the YIG coupling independent of the YIG tuning. This must be prevented by minimizing loop inductance or redesigning the oscillator for an inductive input impedance.

The third type of oscillator is the varactor-tuned or voltage-tuned oscillator. A voltage can be used to change an electronic capacitor in the circuit and thus change the resonant frequency. The advantage of this technique is speed in tuning, typically 1 μsec or faster. The disadvantages are the low Q of the varactor resonator, the posttuning frequency drift, which is usually attributed to the varactor, tuning nonlinearity, narrow bandwidth (usually less than an octave), and higher FM noise close to the carrier.

A typical broadband varactor-tuned oscillator is shown in Fig. 4.35. The emitter circuit is nearly an open circuit, and the resonance occurs between the collector-base capacitance in series with the varactor capacitance and the load inductance. The advantage of this circuit is that a series varactor is tuning the relatively large transistor capacitance and thus controlling the frequency over a broad range. Also, the collector of the bipolar is shorted to ground for heat sinking.

The broadband varactor-tuned oscillator can be expanded to two transistors as shown in Fig. 4.36. This circuit will operate in the fundamental mode if inductively coupled at the output and in the doubling mode if capacitively coupled at the output. The high impedance in the emitter provides the negative resistance at the base-collector port.

In the doubling mode, this circuit is called push-push. On one half-cycle Q_1 operates with a pulse of current; on the other half-cycle Q_2 operates with

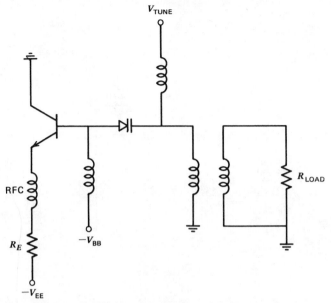

Figure 4.35 Broadband varactor-tuned oscillator.

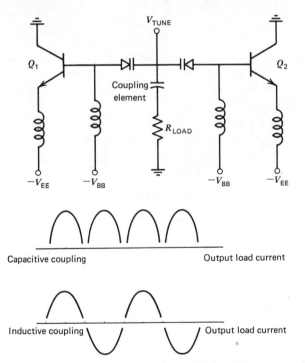

Figure 4.36 Broadband varactor-tuned oscillator with two transistors.

179

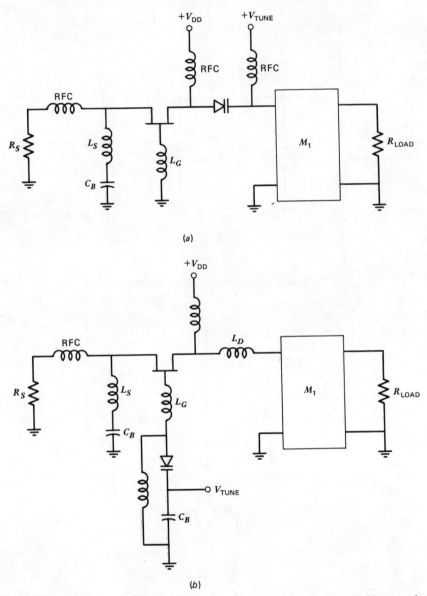

Figure 4.37 GaAs MESFET VCO design. (*a*) Varactor in drain; (*b*) Varactor in gate.

a pulse of current. Thus the double frequency or second harmonic is very strong at the load. In the fundamental mode, this is a push-pull circuit. Notice that the power output is twice the power of a single bipolar transistor. This type of circuit can also be used for tripling with reduced output power by using different varactors at the two transistors. The name "osciplier" (oscillator-multiplier) is used to describe this type of circuit.

Some practical GaAs MESFET varactor-tuned oscillators are shown in Fig. 4.37. The varactor may be placed in series with the drain circuit or in series with the gate circuit. Both oscillators take power out of the drain. When a medium-power FET with $Z = 1200$ μm is used, these oscillators give 50 to 200 mW of output power over 8 to 11.5 GHz (Fig. 4.37a) or 8 to 13 GHz (Fig. 4.37b).

The design of a wideband, varactor-tuned GaAs MESFET oscillator over 9 to 18 GHz using a 300-μm device proceeds as follows (see Fig. 4.38): Choose a suitable load impedance for low-impedance loading at high frequency. Investigate the angle of S_{11} versus frequency. An inductive reactance is required in the gate lead to make S_{11} greater than unity. An open stub will resonate the device at 17 GHz with $L_s = 0.2$ nH. The same stub at 11 GHz requires $L_s = 1.2$ nH. The varactor and series inductance must be designed to fit this oscillation condition. An experimental oscillator of this configuration has achieved 10 mW at 11 to 17 GHz.

In summary, practical oscillator designs can be reduced to fixed-tuned, magnetically tuned, and voltage-tuned designs. The system requirements dictate the type of oscillator required. The design can be accomplished with small-signal S-parameters, which will correctly predict the frequency of oscillation. The output power is determined by the dc bias condition and the possible power saturation in the circuit. The name of an oscillator circuit is usually determined by the type of feedback and the "common mode." Since

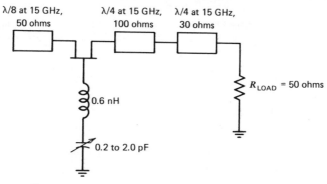

Figure 4.38 Ku band voltage-controlled oscillator.

the circuit can often be drawn with the common mode chosen two different ways, the terminal that delivers output power to the load uniquely determines the oscillator name.

Bibliography

4.1 H. Abe, Y. Takayana, A. Higashisaka, and H. Takamizawa, "A Highly Stabilized Low-Noise GaAs FET Integrated Oscillator with a Dielectric Resonator in the C-Band," *IEEE Transactions on MTT*, Vol. MTT-20, March 1978, pp. 156–162.

4.2 G. D. Alley and H. C. Wang, "An Ultra-Low Noise Microwave Synthesizer," *IEEE Transactions on MTT*, Vol. MTT-27, December 1979, pp. 969–974.

4.3 J. A. Barnes, A. R. Chie, and L. S. Cutter, "Characterization of Frequency Stability," *IEEE Transactions on Instruments and Measuring*, Vol. IM 20 May 1971, pp. 105–120.

4.4 G. R. Basawapatna and R. B. Stancliff, "A Unified Approach to the Design of Wide-Band Microwave Solid-State Oscillators," *IEEE Transactions on MTT*, Vol. MTT-27, May 1979, pp. 379–385.

4.5 J. Gaffuil and J. Caiminade, "Low-Frequency Noise in GaAs Schottky-Gate FETS," *Electronic Letters*, Vol. 10, June 1974, pp. 266–267.

4.6 S. Hamilton, "FM and AM Noise in Microwave Oscillators," *Microwave Journal*, June 1978, pp. 105–109.

4.7 D. J. Healy, III, "Flicker of Frequency and Phase and White Frequency and Phase Fluctuations in Frequency Sources," *Proceedings of the 26th Annual Symposium on Frequency Control*, 1972, pp. 43–49.

4.8 K. Kurokawa, *An Introduction to the Theory of Microwave Circuits*, Academic, New York, 1969, Chapter 9.

4.9 K. Kurokawa, "Noise in Synchronized Oscillators," *IEEE Transactions on MTT*, Vol. MTT-16, April 1968, pp. 234–240.

4.10 K. Kurokawa, "Microwave Solid State Oscillator Circuits," Chapter 5 in M. J. Howes and D. V. Morgan, Eds., *Microwave Devices*, Wiley, New York, 1976.

4.11 D. B. Leeson, "A Simple Model of Feedback Oscillator Noise Spectrum," *Proceedings of the IEEE*, Vol. 54, February 1966, pp. 329–330.

4.12 J. H. Lephoff and P. Ramratan, "FET vs. Bipolar: Which Oscillator is Quieter," *Microwaves*, November 1980, pp. 82–83.

4.13 M. Maeda, K. Kimura, and H. Kodera, "Design and Performance of X-Band Oscillators with GaAs Schottky-Gate Field-Effect Transistors," *IEEE Transactions on MTT*, Vol. MTT-23, August 1975, pp. 661–667.

4.14 E. C. Niehenke and R. D. Hess, "A Microstrip Low-Noise X-Band Voltage Controlled Oscillator," *IEEE Transactions on MTT*, MTT-27, December 1979, pp. 1075–1079.

4.15 P. Ollivier, "Microwave YIG-Tuned Oscillator," *IEEE Journal of Solid-State Circuits*, Vol. SC-7, February 1972, pp. 50–60.

4.16 J. C. Papp and Y. Y. Koyano, "An 8-18 GHz YIG-Tuned FET Oscillator," *IEEE Transactions on MTT*, Vol. MTT-28, July 1980, pp. 762–766.

4.17 A. Podcameni and L. A. Bermudez, "Stabilised Oscillator with Input Dielectric Resonator: Large Signal Design," *Electronic Letters*, Vol. 17, January 1981, pp. 44–45.

4.18 R. M. Rector and G. D. Vendelin, "A 1.0 Watt GaAs MESFET Oscillator At X-Band," *1978 International Microwave Symposium*, June 1978, pp. 145–146.

4.19 R. Rippy, "A New Look at Source Stability," *Microwaves*, August 1976, pp. 42–48.

4.20 D. Scherer, "Learn About Low-Noise Design," *Microwaves*, April 1979, pp. 120–122.

4.21 D. Sodini, A. Touboul, G. Lecoy, and M. Savelli, "Generation-Recombination Noise in the Channel of GaAs Schottky-Gate Field-Effect Transistor," *Electronic Letters*, Vol. 12, January 1976, pp. 42–43.

4.22 A. A. Sweet, "A General Analysis of Noise in Gunn Oscillators," *Proceedings of the IEEE*, August 1972, pp. 999–1000.

4.23 R. J. Trew, "Design Theory for Broadband YIG-Tuned FET Oscillators," *IEEE Transactions on MTT*, Vol. MTT-27, Jan. 1979, pp. 8–14.

4.24 H. Q. Tserng and H. M. Macksey, "Wide-Band Varactor-Tuned GaAs MESFET Oscillators at X- and Ku-bands," *IEEE MTT-S International Microwave Symposium Digest*, 1977, pp. 267–269.

4.25 W. Wagner, "Oscillator Design by Device Line Measurement," *Microwave Journal*, February 1979, pp. 43–48.

4.26 R. G. Winch and J. L. Matson, "Ku-Band MIC Bipolar VCO," *Electronics Letters*, Vol. 17, April 1981, pp. 296–298.

APPENDIX A

DERIVATION OF
THE STABILITY FACTOR

The derivation of (1.100) was given in Ref. 1.5 by Carson, but the proof is repeated here for completeness.

The graphical solution for stability from Fig. 1.10b is

$$|C_G| - r_G > 1 \tag{A.1}$$

Using (1.97) and (1.98) gives

$$\left| \frac{|S_{22} - DS_{11}^*| - |S_{12}S_{21}|}{|S_{22}|^2 - |D|^2} \right| > 1 \tag{A.2}$$

which becomes

$$\left| |S_{22} - DS_{11}^*| - |S_{12}S_{21}| \right|^2 > \left| |S_{22}|^2 - |D|^2 \right|^2 \tag{A.3}$$

We square and group terms to obtain

$$2|S_{12}S_{21}||S_{22} - DS_{11}^*| < |S_{22} - DS_{11}^*|^2 + |S_{12}S_{21}|^2 - \left| |S_{22}|^2 - |D|^2 \right|^2 \tag{A.4}$$

Using the identity

$$|S_{22} - DS_{11}^*|^2 = |S_{12}S_{21}|^2 + \left(1 - |S_{11}|^2\right)\left(|S_{22}|^2 - |D|^2\right) \tag{A.5}$$

we square (A.4) and combine terms to give

$$\left(|S_{22}|^2 - |D|^2\right)^2 \left\{ \left[\left(1 - |S_{11}|^2\right) - \left(|S_{22}|^2 - |D|^2\right) \right]^2 - 4|S_{12}S_{21}|^2 \right\} > 0 \tag{A.6}$$

which finally results in

$$2|S_{12}S_{21}| < 1 - |S_{11}|^2 - |S_{22}|^2 + |D|^2 \qquad (A.7)$$

$$k = \frac{1 - |S_{11}|^2 - |S_{22}|^2 + |D|^2}{2|S_{12}||S_{21}|} > 1 \qquad (A.8)$$

This is the stability factor used to describe the unconditional stability of a two-port.

APPENDIX B

AN IMPORTANT PROOF CONCERNING STABILITY

It is not obvious that the inequality

$$\frac{B_1}{2|C_1|} > 1 \tag{B.1}$$

implies that the stability factor k is greater than unity. The proof follows.

Substituting the definitions of B_1 and C_1 from Eqns. (1.118) and (1.117) into (B.1) gives

$$\frac{1 + |S_{11}|^2 - |S_{22}|^2 - |D|^2}{2|S_{11} - S_{22}^* D|} > 1 \tag{B.2}$$

$$1 + |S_{11}|^2 - |S_{22}|^2 - |D|^2 > 2|S_{11} - S_{22}^* D| \tag{B.3}$$

Using the identity

$$|S_{11} - S_{22}^* D|^2 = |S_{12} S_{21}|^2 + \left(1 - |S_{22}|^2\right)\left(|S_{11}|^2 - |D|^2\right) \tag{B.4}$$

and squaring both sides of (B.3) gives

$$\left(1 - |S_{22}|^2 + |S_{11}|^2 - |D|^2\right)^2 > 4|S_{12} S_{21}|^2 + 4\left(1 - |S_{22}|^2\right)\left(|S_{11}|^2 - |D|^2\right) \tag{B.5}$$

Rearranging terms yields

$$\left[\left(1 - |S_{22}|^2\right) + \left(|S_{11}|^2 - |D|^2\right)\right]^2 - 4\left(1 - |S_{22}|^2\right)\left(|S_{11}|^2 - |D|^2\right)$$
$$> 4|S_{12} S_{21}|^2 \tag{B.6}$$

Now using the identity

$$(a+b)^2 - 4ab = (a-b)^2 \tag{B.7}$$

will give

$$\left[(1 - |S_{22}|^2) - (|S_{11}|^2 - |D|^2)\right]^2 > 4|S_{12}S_{21}|^2 \tag{B.8}$$

which can be written

$$\left[(1 - |S_{11}|^2 - |S_{22}|^2 + |D|^2)\right]^2 > 4|S_{12}S_{21}|^2 \tag{B.9}$$

Taking the square root of both sides yields

$$1 - |S_{11}|^2 - |S_{22}|^2 + |D|^2 > 2|S_{12}S_{21}| \tag{B.10}$$

which completes the proof.

INDEX